Making Microchips

Making Microchips

Policy, Globalization, and Economic Restructuring
in the Semiconductor Industry

Jan Mazurek

The MIT Press
Cambridge, Massachusetts
London, England

Set in Sabon by Graphic Composition, Inc.
Printed and bound in the United States of America.

Library of Congress Cataloging-in-Publication Data

Mazurek, Jan, 1965–
    Making microchips : policy, globalization, and economic restructuring in the semiconductor industry / Jan Mazurek.
        p.   cm. — (Urban and industrial environments)
    Includes bibliographical references and index.
    ISBN 0-262-13345-8 (alk. paper)
    1. Semiconductor industry—Employees—Health and hygiene. 2. Integrated circuits industry—Employees—Health and hygiene. 3. Computer industry—Employees—Health and hygiene. I. Title. II. Series.
    HD7269.S44M39   1999
    338.4'76213815—dc21   98-35601
                                                                      CIP

in memory of Julie Roqué (1957–1996)

Contents

Foreword

In this excellently researched, readable, informative treatise on what will be one of the most important industries of the 21st century, Jan Mazurek skillfully weaves issues related to innovation and industrial policy, environmental concerns, and discussions of public participation and regulatory reform into a significant contribution to scholarship at the interface of industrial innovation and environmental policy.

Constant and rapid innovation distinguishes the microelectronics industry from other manufacturing industries. Historically, microchip manufacturers have routinely doubled the number of transistors on a silicon wafer about every 18 months. Continuous innovation is sometimes achieved by changing from silicon to more hazardous substrate materials, such as gallium arsenide. From the perspective of the final products, however, the environmental problems associated with microchip manufacture appear relatively minor in comparison to those associated with more traditional industries. Indeed the microelectronics industry is often described as a "clean industry."

From the perspective of worker exposure, however, the constantly shifting sets of chemicals used in manufacturing microchips often present hazards that outstrip those found in other industries. It is estimated that 99.9 percent of the materials used to manufacture chips are not contained in the final products. Further, substances used in manufacture pose both a resource and a disposal challenge. Life-cycle analysis is an increasingly used analytic technique for assessing the environmental impacts of a particular product. Its application routinely leaves out the front end of the manufacturing process—extraction and refining of starting materials—as well as exposure of workers and use of resources. This omission in lifecycle analysis is particularly misleading in the case of microelectronics, which, in addition to exposing workers to some of the most hazardous substances used in any kind of manufacturing, understates both the environmental toxicity and the worker toxicity associated with extraction and fails to highlight the large use of energy and water.

However, *Making Microchips* is mostly not about the hazards and the resource issues associated with the industry. This work challenges the assumptions of policies designed to promote the competitiveness of this dynamic, expanding industry. It reveals how dramatic economic and organizational changes within the industry often confound efforts to track, monitor, and manage pollution. Both representatives of the industry and regulators have argued that development and technological change in this industry are too fast for regulations to keep up with them. Further, they posit that regulatory flexibility is essential and warranted in an industry characterized by economic vigor and technological sophistication and knowledge.

In recent years the US Environmental Protection Agency has designed two voluntary initiatives to accommodate the need to protect the environment and the desire to make the microelectronics industry a centerpiece of improvement in the United States' international competitiveness and of high-tech employment. Project XL was designed to offer regulatory relief on a firm-by-firm basis by offering firms quasi-exemptions from complying with specific regulatory requirements in exchange for their commitment to superior environmental performance, measured as continuous reduction in the use of toxic substances. Public involvement and scrutiny were seen as important checks on possible abuse. Though technically Project XL was not authorized by environmental laws, the implicit argument was that there would be no one to challenge technical violations in court if all the players were on board. There was nothing surreptitious about Project XL. Indeed, agency officials would informally joke that "if it isn't illegal, it isn't XL."

The second voluntary initiative designed to accommodate environmental and growth interests, the Common Sense Initiative (CSI), was to be negotiated sector by sector. Petroleum refining, automobile manufacturing, and computers and electronics were three of the six sectors involved early in the initiative. In those sectors, the stakeholders were expected to negotiate environmental requirements for the sector, and again, even if some of the negotiated agreements were technically illegal, no legal challenges were expected, because all the stakeholders would have been involved.

Neither the Common Sense Initiative nor Project XL has been a resounding success. The CSI for computers and electronics involved only a few large firms, and it has failed to encourage pollution prevention or new environmental technology, focusing instead on product recycling.

As with the Common Sense Initiative, a number of national environmental groups and industry watchdog organizations questioned whether a Project XL agreement involving one of Intel's Arizona facilities provided environmental benefits equal to those that would have been expected under existing federal laws. At first glance, Intel's emissions compare favorably with those of other firms in the chip industry. However, the record of reduction of industry-wide toxic releases looks impressive until examined more closely. It appears that, although the reductions in emissions reported to the Toxics Release Inventory probably are attributable in part to improved environmental performance, other factors are also important.

Today, although most US major chip companies still maintain their headquarters and their planning and design functions in Silicon Valley, chip manufacture occurs in other Western states and, increasingly, abroad, near overseas markets. And "fablessness"—production outside the United States—is increasing. Although US firms still control microprocessor and custom chips, they increasingly contract routine manufacturing out to developing countries, and Asian firms now control the manufacture of memory chips. Reductions in TRI emissions for US-based manufacturing must be seen in the light of this globalization of production. It seems, then, that voluntary environmental initiatives—at least in the microelectronics sector—have failed to address two concerns: (1) whether relaxing regulatory requirements creates sufficient incentives to encourage cleaner technology which could have been brought about by otherwise stringent regulation and (2) whether the current system targets an industry that increasingly locates production, jobs, and pollution abroad. The lack of appropriate government oversight in the context of so-called voluntary initiatives should provide some hard lessons for the future.

Making Microchips gets beyond the rhetoric of regulatory reform and laissez-faire economic development in the context of microelectronics, a crucial industry for the 21st century. There is a great need to come to terms with an increasingly globalized economy without transferring jobs and pollution abroad. Thus, this work is a must-read for those involved in computer and chip technologies and for those involved in regulatory and environmental policy.

Nicholas A. Ashford
Professor of Technology and Policy
Massachusetts Institute of Technology

Acknowledgments

My interest in Silicon Valley was sparked in 1984 during an aborted bus ride from downtown Santa Cruz over the treacherous Highway 17 to the San Jose Airport. The driver missed the airport stop and deposited me somewhere among a series of giant warehouses that appeared to have swallowed a commuter college campus. I know now that those labyrinthine structures were really wafer fabs. I'd like to thank that anonymous bus driver who missed my stop. I also would like to thank Ward Coffey, a Santa Cruz shaper of sail and surfboards, who convinced me that I am a better writer than surfer.

I started to think about the environmental implications of microchip manufacture during a graduate seminar at UCLA. The class was taught by a team of urban planners, engineers, and public health experts. This book evolved from my work with two of them. Robert Gottlieb and the late Julie Roqué have been my biggest advocates and best friends. I deeply regret that Julie is no longer here to guide students who don't come from the best neighborhoods, or to ensure that federal policies protect people who continue to live where they come from. At her memorial service in April 1997, President Clinton's science advisor, John Gibbons, referred to his former staffer as a "rising star." Julie also was someone who liked to run around UCLA with a squirt gun. My thanks also go to UCLA's Michael Storper, who introduced me to economic geography, a discipline that informs this work.

Countless colleagues and associates at two research organizations in Washington helped to supply the book's policy focus. In particular, I am indebted to Richard Minard and DeWitt John at the National Academy of Public Administration and to Terry Davies, director of the Center for

Risk Management at Resources for the Future, for providing me with firsthand exposure to federal environmental policymaking. Although they taught me a great deal, I take full responsibility for both the accuracy of information presented here, as well as the way in which it is interpreted. One of my former colleagues at Resources for the Future recently told me that my journalistic approach to research demonstrated to him for the first time the value of his doctorate in economics. I take that as a compliment.

Perhaps my greatest debt goes to the public servants at the US Environmental Protection Agency, an agency that must constantly balance the conflicting interests of industry, state and local governments, Congress, and nonprofit organizations. My unremitting belief in the EPA's potential brought me to Washington and continues to motivate my research. I also want to thank the numerous representatives from industry and nonprofit organizations who supplied the background information contained herein. In particular, I would like to thank Ted Smith and his team at the Silicon Valley Toxics Coalition for sharing with me their organization's perspectives.

Contrary to popular conception, penning a book is not a solitary exercise but one that relies on the emotional and material support of family and friends. Had I known in advance just what a drain such an undertaking would be on the people I care about, I would have elected to forgo this exercise completely. Thanks in particular to my mom, Ursula Mazurek, and to Patrick Hagan for holding my hand through this long, difficult, and sometimes rewarding process.

Making Microchips

Introduction

Quick or dead
—Intel Corporation motto

In 1996, in an experiment aimed at improving the effectiveness and efficiency of environmental laws, the US Environmental Protection Agency and the Intel Corporation entered into an agreement allowing Intel to make changes in the production of microchips more "quickly" than its foreign competitors. Though no one knows for sure, some Intel managers say that production delays can cost the company millions of dollars in lost revenue each day. In exchange for the ability to respond more rapidly to market signals, Intel pledged to reduce more pollution than federal law requires.

The Intel agreement was struck under Project XL, a pilot program that the Clinton administration calls the "crown jewel" of its efforts to reduce regulatory burdens on industry (Clinton and Gore 1995). As President Bill Clinton put it when he announced the initiative in 1995, Project XL is premised on the idea that "in many cases companies know their business a whole lot better than the Government does" (Clinton 1995). The initiative is designed to encourage companies to demonstrate environmental "excellence and leadership."

Intel's Project XL agreement covers operations at the mammoth "Ocotillo" manufacturing site in Chandler, Arizona, a fast-growing desert suburb near Phoenix. The centerpiece of Intel's XL agreement is an air permit that approves routine production changes before the manufacturer makes them. Microchip manufacturing may require up to 45 process chemistry modifications per year, and multiple equipment changes

(Hatcher 1994, pp. 7–8). Under current federal laws, regulators can require chip manufacturers to obtain approval for each change in chemistry or equipment (US EPA 1996a).

Policy makers praise the Project XL agreement as a way to lower costs and to promote the competitive advantage of US firms that are willing to control pollution more stringently than the laws require. Intel's president, Craig Barrett, was quoted in the 15 December 1997 *Albuquerque Journal* as having stated, during a visit to his company's facilities in Rio Rancho, New Mexico, that Intel would have to build its new factories overseas if the US government and the states were to fail to update their regulatory process to move as rapidly as the industry moves.

Ostensibly, Project XL represents a way to improve regulations and keep manufacturing jobs in the United States. Yet 130 community, environmental, and labor organizations have called Intel's XL agreement with the US Environmental Protection Agency a "sweetheart deal" (CRT 1996a). Watchdog groups such as the Silicon Valley Toxics Coalition charge that Intel's Project XL agreement simply gives "extra leniency" to an industry in which the breathtaking pace and complexity of technical change threaten to outstrip society's ability to understand the environmental effects of high-tech manufacturing.

In 1995, the year that President Clinton unveiled Project XL, several environmental groups, backed by the Jessie Smith Noyes Foundation, a New York philanthropic organization, published a report that was highly critical of Intel's environmental record at a plant in Rio Rancho, New Mexico (SWOP 1995). The publication was preceded by a 1994 shareholder resolution, spearheaded by the Noyes Foundation, which said that Intel "jeopardizes stockholder investments by picking environmentally risky sites for its operations" (Greene 1996, p. 25).

Chip makers received more adverse publicity in 1996 when a group of line workers at an IBM fab[1] in East Fishkill, New York, filed a lawsuit in Manhattan State Supreme Court alleging that the workers' cancers were caused by exposure to chemicals at the plant. These workers received national attention in 1997, when their story was the subject of a segment of the NBC TV news magazine *Dateline*. Industry and trade

---

1. In the parlance of the industry, the facilities in which the silicon wafers from which microchips will be made are fabricated are "fabs."

groups maintain that semiconductor manufacturing is, in general, "clean" (SIA 1997).

The purpose of this book is not to resolve the unremitting technical debates about the risks to humans and the environment posed by chemicals used in the chip industry, but to challenge the assumptions of policies designed to promote the competitiveness of this dynamic and expanding industry. Current federal initiatives seek primarily to reduce the cost of environmental regulation. The traditional assumption is that regulations increase cost. Largely ignored is the degree to which economic change makes it more difficult to adequately regulate and manage pollution. Initiatives such as XL suffer from the lack of an analytic perspective that would consider how "restructuring"[2] changes both how and where companies produce chips. It is not inconceivable that environmental policies may be one day used to promote the competitiveness of US companies. However, current initiatives lack an analytic perspective that would consider how organizational and economic change complicate environmental management.

Moore and More

Ever since its inception, in the 1960s, the semiconductor industry's focus on rapid innovation has confounded attempts to evaluate its effects on the environment and on human health (LaDou 1986). More recently, major shifts in how and where US firms produce chips have made tracking and evaluating the environmental impact of chip manufacturing even more complicated.

In 1965, Gordon Moore, Intel's co-founder, accurately predicted that the industry's focus on continuous innovation would double the power of computer chips every 18–24 months. This phenomenon, which came to be known as Moore's Law, has yielded enormous profits for the industry and has expanded the array of electronics and computer products available to consumers. In 1997, worldwide semiconductor sales came to $137.2 billion (SIA 1998). Sales are expected to climb to $1 trillion by 2005, when chips may well be found in everything from pill bottles

---

2. The use of the word 'restructuring' will be explained in the next section.

that remind patients when to take their medicine to cars that perform their own routine maintenance and know how to navigate (Malone 1996, p. 53).

To manufacture microchips to meet the growing demand, analysts estimate, firms around the world are planning and constructing 127 new fabs at a total cost of more than $115 billion (DeJule 1998). To date, there has been no systematic public or private inventory of the number of fabs in the United States or abroad. According to one industry trade organization, there were more than 900 fabs in 1993. Of those, about 350 were located in the United States (MCC 1993, p. 97).

Since 1994, in the United States alone, at least 24 new fabs have been announced and/or built. According to Sematech, a public-private research consortium, such growth is expected to create approximately 40,000 well-paying manufacturing jobs (Sematech 1997). In addition to expanding the number of plants needed to meet the demand for chips, manufacturers are dramatically modifying their production methods. Chips are made from silicon disks or wafers, which have historically ranged from 4 to 8 inches in diameter. Depending on the wafer's size and the type of chips being produced, 250–400 chips may be made from a single wafer. The manufacturer's objective is to maximize the number of chips per wafer (the "yield"). A high-volume fab may churn out more than 30,000 wafers per month. In 1998 major companies sought to increase chip yield by going to 12-inch wafers—a change that required the use of new inputs, new processes, and new manufacturing equipment.

The industry's clean outward appearance and clean image tend to eclipse its glaringly poor environmental record. A fab looks more like a university or a research park than like a conventional manufacturing plant. In "clean rooms," workers clad in head-to-toe "bunny suits" create chips in an antiseptic environment half a million times more filtered than the average hospital operating room. Yet Moore's Law is largely attributable to the use of some of the most toxic chemicals used in any industry (LaDou 1986; LaDou and Rohm 1998).

Chip manufacturing also is distinguished from some other industries by the degree to which firms have congregated in specific places. The growth and development of the semiconductor is inextricably linked to a region once known as the prune capital of America. Now called Silicon

Valley, this 50-mile stretch of land south of San Francisco houses the highest concentration of semiconductor and electronics companies in the world. In addition, it contains 29 of the US Environmental Protection Agency's "Superfund sites"—the largest collection of toxic contamination spots in the United States (Smith and Woodward 1992).

Though many of the environmental effects of the chemicals used to fabricate wafers remain undocumented, a few of the substances used in fabs have been linked to spontaneous abortions among female fab workers. The industry has announced plans to phase out chemicals that may cause miscarriage and other reproductive health problems, as well as those that contribute to ozone deterioration. Some health experts believe that chemicals used to manufacture microchips can cause cancer. Because it can take cancer up to 30 years after exposure to become manifest, some believe that the health effects of working in wafer fabs are just starting to come to light. The use of certain solvents contributes to fabs' emissions of exotic and hazardous air pollutants.

In addition to toxic chemicals, microchip manufacturing calls for enormous amounts of water to ensure that microscopic circuits are free of contaminants, the leading cause of low yields and product failures. Environmental and equity issues related to the industry's water demands are compounded by the shift of manufacturing facilities away from Silicon Valley to arid regions of the American Southwest.

Restructuring Defined

Though industry dynamism has always posed difficulties for environmental managers, the recent bid by US microchip firms to regain market dominance further complicates efforts to track and assess the environmental effects of chip manufacturing. Since the late 1980s, major US chip firms have improved their competitiveness by dramatically changing how, and sometimes where, they manufacture chips. The term they have used for this is "restructuring."

Design and manufacturing often no longer occur under the same company's roof, as was once the norm. In some cases, US firms and their former competitors (predominantly foreign) jointly construct and operate billion-dollar chip plants. In other cases, chip companies have simply shuttered costly fabs. While still considered chip companies, "fabless" firms contract chip manufacturing out to "foundries"—firms that focus primarily on manufacturing rather than on research, design, and sales. Some foundries are found in the United States, but many are located in Japan, in Korea, and now in Taiwan and Singapore.

A handful of companies that make microprocessors and other sophisticated microchip products continue to construct and operate fabs in order to maintain control over manufacturing processes. However, in response to competitive pressures, chip makers have refined their planning and design strategies to the point that they are able to build new fabs hundreds and sometimes thousands of miles from Silicon Valley. By 1994, Texas had supplanted California as the capital of wafer fabrication. The ability of US firms to locate new fabs in foreign markets may improve these firms' economic competitiveness.

By some measures, the environmental performance of fabs located in the United States has improved. According to the Toxics Release Inventory (TRI),[3] releases to the environment and transfers to treatment facilities from semiconductor firms declined by about 3 percent from 1988 to 1995, the last year for which figures are available (US EPA 1997b). Some of these improvements are attributable to the industry's aggressive efforts to reduce pollution. However, they may also be due in part to the fact that firms operating in the United States are not required to report emissions from manufacturing sites abroad to the EPA.

Does Manufacturing Matter?

The sweeping changes in how and where semiconductor manufacturers organize production mirror changes in the advanced capitalist economies since the late 1960s. Observers generally agree that these economies have witnessed a gradual erosion of high-paying manufacturing jobs and an attendant rise in the number of service jobs. However, there is little agreement on what caused these changes, or on what is the appropriate policy response.

3. The TRI is a US Environmental Protection Agency database that records the results of the required annual reports on releases and off-site transfers of 643 toxic chemicals.

Many observers maintain that "de-industrialization" is the result of intensified global competition and shrinking profit margins (Harrison and Bluestone 1982). Lester Thurow (1992) and like-minded economists attribute much of the decline of heavy industry in the United States to the resurgence of other advanced capitalist economies after World War II and to the economic emergence of Singapore, Taiwan, and other "developing countries." They argue that, owing to economic shifts, US manufacturers no longer maintain unbridled hegemony and are consequently forced to seek new methods to reduce costs and promote innovation. In contrast, the economist Paul Krugman (1996) maintains that declines in US manufacturing are largely driven by technological change and by domestic shifts in labor productivity. Though they may appear diametrically opposed, both accounts help to explain restructuring in the US semiconductor industry.

The idea that foreign countries pose the primary threat to US chip manufacturing jobs has tended to color the US government's policies since the mid 1980s, and particularly under the Clinton administration. Though Congress has consistently funded efforts to support the microchip industry, efforts to help the industry regain market leadership during the Bush administration were characterized by the industry as "lukewarm" (Leopold 1994, p. 10). In 1992, President George Bush signed a military appropriations bill that earmarked $100 million for Sematech. Congress first appropriated matching funds for the public-private consortium in 1987. Sematech was established to promote the competitiveness of US microelectronics manufacturers. Its members include ten of the largest microchip companies. Until 1997, when the federal government stopped funding Sematech, the consortium's corporate members were required to match federal funds. Sematech now obtains all its funding from private sources. Attached to the 1992 appropriations bill was a rider that required Sematech to spend $10 million on developing environmentally safe processes for manufacturing semiconductors. Sematech's plan to develop technology to make chips smaller and faster (SIA 1994) contains a set of qualitative environmental goals.

Since 1992, the Clinton administration has advanced a number of additional strategies to promote the semiconductor and computer industries. Calling leadership in semiconductors critical to the economy, Vice-

President Al Gore has sponsored a number of related government initiatives to build upon the industry's successful recent bid to regain global market share. For example, in 1994 the administration pledged to spend $50 million over 5 years to help improve the manufacturing methods of US chip makers. The industry matched the federal government's contribution, bringing the total to $100 million. That year, the administration also established the Semiconductor Technology Council, an umbrella group with members from industry, government, and academia. In addition to funding research to improve manufacturing methods, in 1996 the Clinton administration successfully pushed a new pact to promote more favorable terms of trade with Japanese firms, which had successfully challenged US chip makers during the 1980s by selling chips below cost.

The aforementioned policies appear to be working. By 1997, semiconductors and computers were the driving force behind the United States' industrial growth. In a special supplement devoted to Silicon Valley, *The Economist* (1997, p. 1) reported that the United States' computer and information technology firms now accounted for at least 10 percent of the gross domestic product. In a similar special issue, *Business Week* (1997, p. 64) reported that in just 5 years computer and semiconductor production had mushroomed to account for 45 percent of US industrial growth. Perhaps the ultimate testament to the microchip's role in the world economy was *Time*'s decision to name Intel's co-founder, chairman, and former chief executive officer, Andrew Grove, its 1997 Man of the Year, calling him "the person most responsible for the amazing growth in the power and innovative potential of microchips" (Isaacson 1997, p. 8).

Largely absent from federal trade policies and from the glowing press reports are an account of the chip industry's environmental legacy and a forecast of what its explosive new growth may portend for human health and for the environment. For example, though *Business Week*'s Silicon Valley supplement lists high housing costs, labor shortages, and traffic jams as some of high technology's downsides, it fails to mention the Valley's dubious distinction as the nation's capital of Superfund sites. Also absent are accounts of studies that document reproductive problems in female fab workers.

The lingering perception that chip manufacturing is clean also is manifest in federal environmental initiatives such as Project XL, which is designed to reward facilities that have spotless compliance records. In contrast to Project XL's facility-based focus, the EPA's Common Sense Initiative targets pollution that is unique to the computer and electronics industries. Both initiatives are designed to reduce regulatory requirements bearing on chip makers and on other high-tech firms.

American chip manufacturers' complaints that environmental regulations harm their competitiveness probably are correct but highly exaggerated. Debates between environmental groups and policy proponents regarding whether or not companies such as Intel should receive relief from environmental regulations will remain deadlocked until environmental managers develop a better framework for assessing the effects of economic restructuring on US chip makers. The Toxics Release Inventory helps regulators and interested citizens to evaluate some emissions from facilities owned and operated by one company in one place, but it does not track emissions due to fabless companies, nor does it promote accountability in cases where fabs are owned and operated by multiple firms.

The industry's ability to build wafer fabs anywhere also may impair the US government's efforts to balance environmental and economic concerns. The promise of creating manufacturing jobs gives chip firms enormous economic leverage over states, localities, and developing countries eager to lure jobs and revenue. Often, the governments most eager to lure chip manufacturers are the governments most lacking in resources and institutions with which to assess the environmental challenges of microchip manufacturing.

Regulatory Reform

Concomitant with economic change is a movement to modify how federal laws protect air, water, and land. Though few would argue that the current environmental regulatory system is perfect, few agree on what is broken or on how to fix it (Davies and Mazurek 1997, p. 48).

Since the significant expansion of federal environmental laws during the late 1960s, environmental economists have conducted a number of studies on various industries. The results of these studies generally reinforce the idea that laws designed to control pollution tend to raise manufacturing costs and lower productivity.

Economists generally conclude that the most effective and efficient way to improve environmental laws is to develop policies that harness market forces. Such instruments include "emissions trading" programs that allow firms to decide how to curb pollution (Kneese et al. 1970; Barbera and McConnell 1990). More recently, the National Academy of Public Administration (NAPA 1995) and Pedersen (1995) have asserted that air, water, waste, and toxic laws are largely victims of their own success. They contend that large companies now have the resources and the environmental personnel to control and monitor pollution better than most government agencies.

Many Fortune 500 companies maintain that the incremental cost of controlling smaller and smaller amounts of pollution fails to justify additional regulations (Davies and Mazurek 1996). A number of companies recommend that Congress relax laws that target large industrial sources and focus on more difficult pollution sources, such as land use and commuting patterns that contribute to lingering air and water woes. Others companies, most notably Intel, would prefer to see a lowering of the costs associated with environmental permits—permits that may cause costly manufacturing delays.

Still others maintain that environmental regulations are based on over-conservative scientific assumptions about risks to humans and the environment. The 104th Congress faulted the technical methods used by the EPA and other federal agencies to assess environmental risk. Risk assessments that inform regulatory standards are based on complex models that yield uncertain results. Depending on the assumptions, these models may show that a chemical may harm a few sensitive individuals, hundreds, or thousands.

Models must be used to assess the effects of chemicals because of the obvious ethical and practical limits on human trials. Complex, quantitative exposure-assessment models generate probabilistic estimates of the effects on humans of exposure to hazardous substances. However, the models require analysts to use a number of simplifying assumptions regarding the person exposed to a particular substance and the length of the exposure. Though it is possible to evaluate the effects of single substance, more realistic assessments of how combinations of chemicals affect humans and the environment simply aren't scientifically feasible.

Some lawmakers claim that overconservative assumptions are responsible for the rising cost and complexity of environmental regulations. They argue that faulty risk assessments lead to overrestrictive and hence costly regulations. For example, a Superfund cleanup decision based on the assumption that children *eat* contaminated soil is likely to require more measures to restrict access and increase soil removal than a decision based on the assumption that children *play on top of* contaminated soil. Between 1994 and 1995, legislators seeking to correct such perceived problems advanced 20 proposals to require agencies to balance the costs of measures designed to protect human health and the environment against the benefits of tighter regulatory standards (Schierow 1994, p. 253). Although such proposals sound reasonable in theory, they assume that science and economics have achieved levels of precision that they have not in fact achieved. A risk assessment typically fails to generate a single precise estimate regarding how much of a substance is harmful; rather, it generates a range of possibilities. In practice, most of the congressional proposals would hamstring regulatory agencies; they might even reduce the effectiveness of earlier environmental laws by prescribing what types of assessment methods scientists should use and what results they should achieve.

Project XL is the Clinton administration's response to the Republican-led efforts to change how agencies assess risk and develop regulations. As envisioned, Project XL would preserve the current statutes by allowing "excellent" firms to violate regulations in exchange for superior environmental results. The Intel agreement is one of about 40 proposed Project XL experiments that would allow greater regulatory flexibility. Their architects originally hoped that such policies would serve better than regulatory reform to simultaneously improve environmental quality and the competitiveness of US companies (Clinton and Gore 1995).

Project XL has fallen short of expectations in part because the EPA lacks the legal authority to let firms operate outside existing pollution control laws (Ginsberg and Cummis 1996). Firms obviously are unwilling to undertake projects that would increase their exposure to EPA enforcement actions and to citizen-led lawsuits. Lack of legal authorization weakens implementation because the EPA is required to divert money and personnel to run Project XL away from programs required by law (NAPA 1995).

However, the semiconductor industry illustrates that efforts to simultaneously improve competitiveness and environmental performance require the EPA to develop far more information about the firm's research methods, market structure, chemical use, and control methods than current laws require. Because chips' value is derived less from the raw materials used to make them than from how they are designed and manufactured, many firms in the chip industry are reluctant to release information of the kind that strategies such as Project XL would require.

In addition to data on pollution and on production, improving how laws and regulations address the semiconductor industry requires environmental regulators to understand why, in recent years, some chip firms have moved their production from Silicon Valley to regions where people are less familiar with the environmental impacts of the industry.

Regulatory experiments such as Project XL and the Common Sense Initiative are designed to reduce the role of government by promoting greater public oversight. Unfortunately, the industry's rapid innovation tempo and the information intensiveness of chip manufacturing confound public efforts to assess and monitor how well regulatory experiments really work.

Black-Box Production

At the simplest level of economic analysis used to guide policy, production takes place in a timeless, placeless ether under a single factory roof. Though firms may differ in the amounts of inputs used and outputs generated, they all accept the same market-determined price for their products. For the sake of simplicity, thorny real-world issues such as adaptation over time, market imperfections, and geography are assumed away.

In most cases, the simple theory of the firm, which involves an optimal input mix and an optimal level of output, provides a perfectly adequate guide for assessing the effects of a proposed environmental policy. Firms discharge pollution to the environment free of cost until regulations or market incentives force them to "internalize" its cost. Regulations that limit how much pollution a firm may emit therefore raise the cost of production.

In *An Evolutionary Theory of Economic Change,* the economists Richard Nelson and Sidney Winter (1982) observe that the theory of the firm advanced in introductory microeconomics is less appropriate as a guide to policy decisions that involve rapidly changing industries. Since *An Evolutionary Theory of Economic Change* was first published, mainstream economists have developed more sophisticated models that attempt to portray how firms adapt to signals over time. For example, Giovanni Dosi (1988) uses a dynamic perspective to examine how the microelectronics industry's focus on continuous innovation distinguishes it from other industries. Until recently, however, a dynamic perspective has not been used to evaluate how environmental regulations affect firms, industries, and consumers (Boyd et al. 1998).

Perhaps more in chip manufacturing than in most industries, time influences how markets are organized. Since the 1980s, the branch of economics known as industrial organization has developed a number of models and empirical studies to better illustrate the composition of markets. Theorists have applied such insights to explain why some producers of products such as memory chips are technological leaders who develop and sell chips first and others are followers who sell chips at lower cost months and sometimes years later (Gruber 1994). Although such insights could help environmental managers better understand Intel's claim, to date they have been used primarily to inform trade policies rather than environmental policies (Tyson 1992; Borrus 1988).

Though advanced economic models are increasingly able to account for time and market structure, few have tackled the more formidable question of how place may be related to production. It is difficult to describe the origin and the growth of the semiconductor industry without referring to where most of the industry's early development occurred. That geography may play a role in economic development has not gone unnoticed by economists, but formal models are lacking. As Krugman (1991, pp. 6–7) observes, the scarcity of models is likely to be due not to a lack of interest but to the fact that economists tend to avoid what they are not able to formalize.

Regional Planning

Most of the work that examines how place contributes to innovation and growth has occurred outside economics, in the fields of geography and regional planning (Angel 1994; Saxenian 1994; Storper and Walker 1989; Scott 1988). In the United States, regional theory dates back to the early days of the Tennessee Valley Authority. However, the perspectives that inform this work are more recent in origin.

The contemporary study of regional planning was developed in response to efforts in the United States and abroad to grow more "Silicon Valleys." Traditional planning theory taught municipal planners and economic development directors to lure firms with simple microeconomic incentives: cheaper land, labor, and capital. Development directors, in contrast, tried to lure high-tech firms by building lavish research parks and by offering tax breaks, worker training packages, and pledges to streamline the environmental permitting process (Miller and Cote 1987). While some regions (such as the Southwest) were successful at luring new chip manufacturing facilities, other regions were saddled with large amounts of vacant industrial space. For example, the state of Illinois invested $128 million in 76 companies between 1985 and 1993 (*Business Week* 1997, p. 138) and experienced only limited success before fiscal constraints caused the development efforts to be abandoned. Similarly, the state of North Carolina and the local counties spent $35 million on facilities for Research Triangle Park and have little to show for the investment so far. Recently, the states of Virginia and Delaware have also attempted to lure chip manufacturers.

The failure of "factor cost" approaches alone to spark high-tech development, coupled with two decades of flagging economic performance in advanced capitalist economies such as Japan and Germany, prompted geographers and planners to study Silicon Valley, the region that incubated most of the major semiconductor breakthroughs and continues to support the world's highest concentration of chip and other high-tech firms. A set of related accounts of why firms cluster, grow, and sometimes leave places such as Silicon Valley have unfolded from these studies.

Insights from economic geography, which have been used to refine economic development strategies, also may help environmental managers to make better decisions and policies. In contrast to simple economic models, economic geography is influenced by theories that illustrate not how economies stabilize but instead how they change over time. Rooted in Joseph Schumpeter's 1934 treatise *The Theory of Economic Development*, economic geography is well suited to charting the economic and environmental effects of semiconductor manufacturing because it examines how technological change can create temporary economic upheavals or "gales of creative destruction." Some analysts use Schumpeter's insights when explaining the growth of service jobs and the attendant decline of heavy industry in Western countries since the late 1960s (Harrison and Bluestone 1982; Piore and Sabel 1984).

Economic geography draws from three distinct but related intellectual strands: transactions cost analysis, flexible specialization, and what the geographer David Angel (1994) refers to as "manufacturing systems analysis." Combined, these theories help to illustrate why current environmental policies fail to account for the effects of restructuring in the semiconductor industry.

Flexible specialization best characterizes the recent trend among some chip firms to get out of manufacturing altogether. Michael Piore and Charles Sabel (1984), who derive their theory from Alfred Marshall's work on nineteenth-century industrial districts, view industrial restructuring as a response to the high, fixed capital costs that characterize vertically integrated mass manufacturing. Piore and Sabel maintain that product cycles are too short and that technology changes too quickly for large integrated firms to respond effectively, and that flexible firms are better able to adapt to market swings than vertically integrated manufacturers with long, stable production runs. They posit that successful firms have responded to intensified global competition by moving from integrated mass production to less centralized forms of manufacturing that require less capital investment and allow relatively short production runs of highly variable, customized products.

Transactions cost analysis (Williamson 1975) helps to explain how places such as Silicon Valley attract certain industries and promote innovation. Transactions cost analysis posits that proximity creates "positive externalities" by reducing some direct costs and some less tangible costs (Scott 1988). For example, Silicon Valley's cluster of high-tech firms enhances transaction opportunities and productivity gains by pooling infrastructure and less tangible know-how. Firms have an incentive to continue to locate there because the region has developed a readily available set of inputs, including a skilled workforce, chemical suppliers, and waste disposal contractors. Once established, the labor and service base creates incentives for firms such as semiconductor manufacturers to remain close to specialized inputs and suppliers that are not available elsewhere.

To the degree that transactions cost analysis helps explain why firms cluster, it also helps to illustrate why industrial clusters break apart ("de-agglomerate," in academic parlance). Transactions cost analysis may, therefore, facilitate prediction of where semiconductor producers will relocate manufacturing facilities. Recent insights into the importance of regional institutions and cultures by the geographers Michael Storper, Richard Walker, and AnnaLee Saxenian help to further explain how some regions lure and keep chip firms and others fail to do so. (See Storper and Walker 1989; Saxenian 1981, 1994.)

Though Piore and Sabel's work helps to illustrate why some firms have moved away from manufacturing homogeneous products, it fails to describe fully why and how the industry has reconfigured from a set of large and small producers to an interlocked system of designers and manufacturers. The industry now comprises small, "fabless" chip companies along with relatively integrated manufacturers such as Intel. The latter, in turn, may have technology agreements, alliances, and foundry partnerships with fabless companies. Contrary to Piore and Sabel's account, the logic of alliances, informal agreements, and foundry partnerships is not merely to reduce capital costs or to design custom chips for niche markets. David Angel's work helps to explain why the focus of this dynamic industry is, more than ever, on moving new products to market.

According to Angel, the United States' technological dominance in semiconductor manufacturing helped to mask serious manufacturing weakness throughout much of the US industry's history. Not until the 1980s, when the Japanese cut into their profits, did US producers identify and correct their substandard manufacturing methods. It is in the context of restructuring that consortia such as Sematech and efforts such as Project XL and the Common Sense Initiative have been developed. Since product quality was improved, Angel has found, the focus of restructuring in the industry is now on shortening the time between product development and shipment while maintaining manufacturing excellence.

Transactions cost analysis, flexible specialization, and manufacturing systems analysis complement the new approaches to environmental management that emerged during the same period. Like the new theories of regional development, the new environmental strategies, known variously as "pollution prevention," "toxics use reduction," and "clean production," require a more detailed understanding of what goes on under the factory roof than policies based on simple microeconomic models.

Though the goals and objectives of the new environmental management approaches vary, they all seek to prevent pollution rather than to control it. Such a perspective requires analysts to focus not only on control and abatement technologies but also on problems all the way up the supply chain to materials extraction and product planning. To conserve materials, for example, analysts must understand why certain substances are used to make products and must identify suitable substitutes. Another method may require assessing product planning and design decisions so as to evaluate where hazardous materials might be removed from the production process (Gottlieb 1995).

The rapidly changing semiconductor industry is thought by some prevention experts and some environmental managers to be a prime candidate for prevention approaches because, as a result of the introduction of new equipment and processes, it ostensibly faces less economic upheaval than other, more stable industries. However, verifying the extent to which the chip makers are likely to reduce their use of toxic substances and prevent pollution is complicated by the fact that, since the chief value of a chip is derived from manufacturing and design know-how, information to aid prevention efforts may be more difficult to obtain from firms in this industry than it is to get from firms in other industries.

Combined, emerging regional theories and toxics use reduction approaches help to illustrate how organizational and geographic change among semiconductor producers complicate policies to improve the regulation of pollution from microchip manufacturing.

Assessing the Environmental Impacts of Economic Change

It is not a coincidence that the "crown jewel" of current reform attempts targets the semiconductor industry. Akin to the way in which the internal combustion engine changed not only how people travel but also where they live, shop, work, and play, the semiconductor has transformed not only how electronic products operate but also how people tabulate and—increasingly—how they communicate. Today, for every motor or electronic device there is at least one semiconductor circuit. It is estimated that as many as 200 billion chips are currently in use (Malone 1996, p. 53), and the number of chips manufactured and sold is expected to skyrocket as the economies of China and other countries continue to develop.

The dynamic nature of the microchip industry delivers substantial economic benefits not only to the giant chip makers but also to computer manufacturers and software companies that design products based on chips. It also benefits users of the Internet and authors who use computers to research and write about chips.

In the United States, data on toxic emissions and off-site transfers lend the chip industry a clean appearance. However, the numbers say little about why emissions and transfers have been declining, where toxics are being emitted, and what firms are responsible. Furthermore, aggregate chemical use and emissions are likely to climb for at least 10 years as an industry that relies on chemicals and water adds production capacity. However, it is questionable whether the mammoth manufacturing facilities now under construction will still be in operation in 2020. Today's industrial greenfields may be tomorrow's Superfund sites or brownfields.

Unfortunately, the policies being used to promote the competitiveness of US firms are incomplete. Project XL is a glaring example. Operating under intense political pressure, the EPA approved Intel's bid to receive regulatory flexibility. The EPA approved the project despite the shortage of data relevant to whether Intel merits such flexibility and whether it is complying with the terms of its XL commitment. Furthermore, though Intel will undoubtedly benefit from the policy, it is not known how much the arrangement will cost or benefit nearby residents, plant workers, consumers, or Intel's corporate rivals.

Lingering questions as to which companies truly merit regulatory flexibility, and questions about the consequences of such initiatives for competitors and consumers, underscore the need for a better framework for assessing the configuration of the semiconductor industry. By examining the experience of the industry, this book provides a starting point for assessing the environmental impacts of the economic changes that have helped to transform how and where microchip makers do business.

1

Smaller, Faster, Better

Those who live by electronics, die by electronics. Sic semper tyrannis.
—Kurt Vonnegut Jr., *Player Piano* (1952)

In *Player Piano,* Vonnegut forecasts a future in which humans, displaced by machines, make work for themselves by building Byzantine bureaucracies. Some beleaguered microchip manufacturers whose profitability depends on their being first to market with new products may feel that this is happening already. Cumbersome regulatory requirements seem to stall production and seem to do little to advance environmental quality.

Because rapid innovation affects the profitability of some chip firms more than that of others, policies to reduce the delays caused by regulation may benefit some firms more than others. Creating environmental policies to target microchip manufacturers requires knowledge of how the industry is structured and of how increasing international competition is changing how individual firms and markets are organized. In order to understand the structure of individual microchip companies and the markets in which they operate, it is important to know something about how and where engineers historically have designed and made microchips.

Silicon Switches

The digital clock, the VCR timer, and even telephone switching equipment rely on electronic pulses controlled and amplified by semiconductors. There are four main classes of semiconductors: integrated circuits (microchips), discrete devices, analog devices, and optoelectronics.[1] Microchips combine hundreds of thousands of transistors on a piece of silicon or, less commonly, gallium arsenide. In 1996, microchips made up 74 percent of the global semiconductor market, followed by discrete devices (13 percent), analog devices (12 percent), and optoelectronics (1 percent) (*Hoover's Online* 1998). In 1970, few people had ever heard of microchips; in 1997, chip companies produced a quintillion transistors—about the number of ants on the planet.

In essence, a semiconductor is merely a switch to turn electric currents on and off. "Off" is represented by 0, "on" by 1. The zeroes and ones—known as "bits," short for binary digits—are the smallest units of information that a machine can recognize. Elaborate strings of zeroes and ones are the underpinnings of the software that instructs computers.

Microchips may be understood as the latest in a long line of human inventions for controlling motion. Examples of early switches include gates to corral livestock or control irrigation.

Shortly after 1900, scientists hit upon the idea of using electricity rather than mechanical power to throw switches. Initially, such operations were possible only in a vacuum. The vacuum tube was the chip's grandparent. ENIAC, the world's first computer, contained 18,000 vacuum tubes, occupied 3000 cubic feet, weighed 30 tons, and required 150 kilowatts of power. It is now outmatched in power and speed by most rudimentary pocket calculator.

The precursor of today's microchip was developed in December 1947 at the Bell Telephone Laboratories in New Jersey by Walter Brattain, John Bardeen, and William Shockley. The trio designed a solid-state amplifier, or transistor, that made it possible to electronically throw switches outside a vacuum on a solid-state device.

In 1954, hoping to capitalize on the semiconductor, Shockley assembled a team of eight engineers and set up shop in Palo Alto. That venture was short-lived because Shockley was more a scientist than a businessman. In 1957, in what has now become Silicon Valley lore, eight engineers

1. Discrete devices are used as electronic switches. Analog devices are used in medical and industrial equipment, automobiles, communications devices, and video equipment. Optoelectronics produce light and are used for electronic displays and data transmission.

left Shockley Semiconductor Labs to found Fairchild Semiconductor. (Shockley called them "traitorous.") Fairchild, in turn, was the progenitor of most latter-day Silicon Valley chip firms, including Intel Corp.

Silicon Valley

"Silicon Valley" is a 50-mile stretch of former farmland that extends from just south of San Francisco Bay to the city of San Jose. Among the other cities that fall within it are Palo Alto, Menlo Park, Mountain View, Milpitas, and Santa Clara. Silicon Valley, not an official geographic designation, is roughly coterminous with Santa Clara County, and statistics for that county serve best to indicate the state of its economy and its environment.

Silicon Valley houses the highest concentration of semiconductor companies in the United States. According to the most recent county-level data compiled by the US Bureau of the Census, 163 Santa Clara County companies designed, made, or sold semiconductors (including integrated circuits, optoelectronics, analog devices, and discrete devices) in 1994 (US DoC 1997a). Collectively, semiconductor products are identified by the federal government under the four-digit Standard Industrial Code (SIC) 3674. To better differentiate among the various types of chips manufactured in the United States, annual SIC reports for most years contain product codes with up to seven digits. Integrated circuits (i.e., microchips) are designated by SIC 36741, whereas discrete devices are designated by SIC 36742. Solar cells and photovoltaic modules are designated by SIC 36749.

In addition to the five-digit codes, the Department of Commerce develops seven-digit product codes that make it possible to distinguish among types of microchips. For example, memory chips fall under SIC 36741 designations 19 and 81–84. Microprocessors fall under SIC 36741 designations 36–39. The seven digit codes also denote product type. For example, a microprocessor with an internal data bus of eight bits is designated as SIC 36741 36, and a chip with 32 bits or more is designated as SIC 36741 39.

As the foregoing discussion suggests, each type of microchip may be differentiated further by its speed or by the amount of information it is able to process or store. For example, memory chips, which store information and read software programs, are designated according to the number of bits they contain. Microprocessors combine memory, logic, and certain discrete functions on a single piece of silicon. SIC codes also identify microprocessors by the number of bits they contain. On store shelves, retailers typically price and advertise microprocessors according to their speed as measured in megahertz.[2] For example, a 200-megahertz microprocessor is typically more powerful and expensive than a 166-megahertz chip.

Chip Geography

A question that has preoccupied economic development experts is why the development and manufacturing of semiconductors took root and flourished in California when the first transistor and first computer were invented on the East Coast. Economic geographers largely agree that the semiconductor industry grew up in the West because the product was sufficiently different from anything made before (Saxenian 1994; Storper and Walker 1989). The device required fundamentally different inputs, production processes, worker skills, and technical knowledge than the established East Coast electronics industry was able to provide.

But why Silicon Valley? Though cheap land and sunshine may have helped, economic geographers maintain that the industry took root there because other factors provided a climate ripe for technological innovation.

Saxenian (1994) found that established electronics firms around Route 128 in Massachusetts failed to dominate the microchip industry because they took fewer risks and pursued a rigid approach to production—an approach characterized by free-standing, vertically integrated firms. One result is that Route 128 firms became "locked into" a technological trajectory—for example, by building products to satisfy military specifications. Saxenian concludes that the Route 128 manufacturers' technical rigidity stifled opportunities for experimentation and learning.

Saxenian also found that another important factor in Silicon Valley's development was Stanford University's support of technological experi-

---

2. One megahertz means a million on-off cycles per second.

mentation. For example, in 1938 Frederick Terman, a Stanford professor, encouraged two of his graduate students, William Hewlett and David Packard, to manufacture a device they had created during their graduate studies. Terman offered them generous fellowships and helped them secure a bank loan to commercialize their audio oscillator. The Walt Disney Corporation was the duo's first customer, and Hewlett-Packard became one of the region's first microelectronics giants. Terman also lobbied for federal funding and established a nexus between Stanford and the emerging local microelectronics industry. Other Stanford endeavors included a research institute, a particle accelerator, and a continuing education program for local professionals. Terman's final effort, the Stanford Industrial Park, was among his most notable contributions. The lease carried the stipulation that only companies interested in sharing technology with the university were permitted to locate there.

One factor common to the early Silicon Valley institutions was their mission of furthering basic research that could be applied to practical commercial uses—not merely to persuading firms to fill up vacant office space. Partly as a result of these early efforts, Silicon Valley had an established technical base in microelectronics by the 1970s, when the semiconductor was first commercially applied.

In addition to explaining why microelectronics firms chose to locate in Silicon Valley, economic geographers show how the region helped the industry to grow. As more and more chip and computer firms were established, they simultaneously developed their own specialized inputs, including skilled labor, chemical suppliers, equipment manufacturers, and hazardous-waste handlers. As the industry grew, so did the demand for inputs, which helped to improve and refine the materials and lower the costs of supplies (Scott and Angel 1987). The industry's growth also helped to spread the costs of other indirect assets, such as roads and schools, among companies.

Clusters of similar firms also help to spread and disseminate ideas. Industrial localization enhances the probability that people engaged in the same activities will meet, exchange information, and, like Shockley's "Traitorous Eight," spin off from one firm to market their ideas (Storper and Walker 1989). Once established, positive "externalities" (as they are known in academic parlance) provide companies with a strong incentive to stay in a region rather than incur the risk of moving to a place where technical information, labor, and materials may be in shorter supply.

Firms cluster to maximize the flow of technical information. However, it is in no individual firm's interest to lose information (or employees) to competitors, thus allowing competitors to gain at no cost what the firm paid to develop. The dual nature of information may be the greatest paradox associated with this new form of manufacturing.

More and Moore Microchips

Beginning in the late 1960s, when US companies (including Intel and Texas Instruments) pioneered microchips, technical breakthroughs in chip design and manufacturing enabled firms to double the number of transistors on a chip roughly every 18 months. Shrinking the space between transistors enhances a chip's speed by reducing the distance a signal must travel. Since 1987, channel widths (distances between circuits) have gone from 10 microns to just 0.35 micron. If this trend continues, they will shrink to less than 0.07 micron by 2011.

One result of unremitting technical breakthroughs is faster, more powerful chips at roughly constant costs. Today's Pentium Pro microprocessor contains 5.5 million transistors; by 2011, Intel's chips may contain at least a billion (Kirkpatrick 1997, p. 63). A microprocessor that sells for $500 today may sell for about $10 in 2002 and about 10 cents in 2007. In 2010, chips equivalent to today's fastest will be cheap enough to provide household objects such as light bulbs with intelligence equal to that of today's desktop computers.

For chip makers, such advances have historically been profitable because the doubling in density was not accompanied by a commensurate jump in manufacturing cost. The alchemy of chip manufacturing is that it combines relatively inexpensive materials such as silicon, oxygen, solvents, and metals to create a product that is worth far more than its weight in raw materials. The value of microchips, therefore, is derived not from the sand, water, and chemicals used to make them, but from the know-how that makes it possible to sandwich transistors side by side into spaces some 200 to 400 times smaller in width than a strand of human hair. As engineers are able to increase the number of transistors on a chip, the average cost per circuit element falls. Traditionally, every chip break-through has halved the manufacturing cost per transistor.

In 1997 Intel released a new chip that shattered the rule, cutting the time needed to double the information stored on a chip to 9 months. The routine improvements in desktop computing to which consumers have grown accustomed are on the verge of dramatic change. Moreover, the very nature of the desktop computer is about to change. In 1997 National Semiconductor, Motorola, and Intel announced plans to develop chips to power network-based personal computers, which are likely to make to-day's desktop machines obsolete.

Manufacturing Uncertainty

For years, Moore's Law helped planners and engineers to predict what types of products to design and in what volume. However, it is harder to predict just how to successfully place 5 million transistors on a thumbnail-size slice of silicon. Unlike products whose sales are based on uniformity, such as Coca-Cola or Big Macs, there is no preset formula for making successive generations of microchips. Instead, each new chip generation is achieved through continual experimentation and refine-ment, or "learning." Fine tuning a product often calls for frequent changes in equipment and in process chemicals.

Intel estimates that roughly one-third of all new microchips require chemistry and equipment that are completely different from those used to make their predecessors. In other words, it is likely that a Pentium II requires inputs entirely different from those used to make a 486. Dur-ing the early stages of a new technology's production cycle, it is nec-essary to make additional modifications in order to improve wafer yield. On average, Intel modifies process chemistry up to 45 times per year and equipment up to 3 times per year (Hatcher 1994, p. 7). As a result of experimentation, microchip output tends to follow a fairly pre-dictable learning curve. At full production, a very large fab may process more than 30,000 8-inch wafers per month. Typically, manu-facturers obtain unfinished silicon ingots from outside suppliers and fabricate them into wafers, which are then sliced into chips. Each wafer may yield 250–400 individual chips, depending on the product and its intended use.

Initially, output, as measured by the number of good chips per wafer, is low as chip makers learn to eliminate duds by modifying chemical processes and equipment configurations. Output then rises over the 2–3-year product cycle, and production costs concomitantly fall. Output falls when the next generation of chips is released, and the cycle starts anew (Dick 1991).

Less-Than-Perfect Competitors

The learning curve suggests that markets for the most technologically sophisticated microchips are not made up of what college economics texts call "perfect competitors": firms of equal size that are all forced to charge the same price. Instead, such markets may consist of a technical leader and followers. The leader bears the cost of designing products and manufacturing methods but is able to set prices and earn monopoly-like profits by releasing chips several months ahead of competitors. Followers appropriate the leader's information and release cheaper chips several months down the road.

The learning curve has been most closely studied in the market segment for dynamic random-access memory (DRAM) chips, the most popular type of random-access memory used in personal computers. Gruber (1992, 1994) demonstrates both theoretically and with firm-level data that the manufacturer that is able to develop and fine tune new DRAM products—to climb the learning curve ahead of its competitors—gains and maintains a dominant market share. Until the followers catch up, the "first mover's" profits are above what the average firm would earn in a perfectly competitive market. The first mover initially sets a new chip's price far above its production cost, on the assumption that only technophiles are willing to pay the price for the most advanced microchip. The first mover experiences monopoly-like profits until the competitors catch up and sell the same chip at a lower price.

If the first mover continually controls the market, what is the incentive for other firms to follow? It is that they are able to learn from the leader's design and manufacturing methods (and mistakes) and are thus spared the time and expense of trial and error. Laggards in the DRAM market can take market share from leaders, but typically only by selling their products well below what it costs to produce them ("dumping"). In theory, at least, it is plausible that a laggard could leapfrog the leader with superior technology. However, outlearning the leader requires a willingness to invest in design and methods, or to incur the costs of product learning.

It is not clear whether the learning-curve effect applies to other segments of the memory market, or to the markets for other products (e.g., microprocessors). For example, anecdotal information shows that Intel controls 80–90 percent of the market for advanced microprocessors by constantly refining chips and by selling smaller, faster versions of its products months and sometimes years ahead of its two chief rivals, Advanced Micro Devices and Cyrix. Restructuring makes it even less clear whether market advantage is due to superior technology.

Limits to Moore's Law

Although the learning-curve effect has long described the structure and the dynamics of some markets, being first may be increasingly insufficient to secure market share. The cost of making smaller, more precise microchips has spurred some firms to question whether the investment in new billion-dollar wafer fabs is worth the return.

Constructing a wafer fab is a mammoth undertaking that typically requires 10 engineers and about 500 contractors. On average, design takes about 6 months and construction about 12. When complete, the average fab contains about a million square feet of space—roughly 36 football fields.

The smallest movements can upset the sensitive manufacturing process, so fabs—and particularly their highly filtered clean rooms—are built to withstand the smallest vibrations. The clean room of Intel's Fab 10, in Ireland, rests on 600 concrete columns set in bedrock. Overall, an entire fab, including the clean-room area, contains enough concrete to build 15 miles of road, and about 3000 miles of reinforcing steel.

Inside, a fab contains about 80 miles of ultra-clean pipe and 150 miles of electrical wire. It has enough cooling capacity and electricity to power 3000 homes. When operational, it consumes enough water (filtered so it is up to 2000 times as pure as tap water) to fill 24 swimming pools each day. The air is filtered continuously. The average fab uses enough compressed air to fill 3000 party balloons per minute.

In 1966 the cost of constructing a fab, measured in current dollars, was about $14 million. By 1996, the price tag was pushing $2 billion and climbing. By 2000, it could be as high as $5 billion. Of this cost, about 75 percent is for fabrication equipment. On average, a large fab has about 250 pieces of equipment, manufactured by about 140 different companies around the world. In recent years the price of the photolithography equipment used to beam smaller and smaller transistor patterns onto silicon has increased about 28 percent per year (Hutcheson and Hutcheson 1996). Because products change so rapidly, state-of-the-art fabs require retooling in less than 5 years; this gives chip makers little incentive to invest in such expensive facilities. Equipment manufacturers appear to be passing their costs on to manufacturers. Despite the rising cost of equipment, most equipment manufacturers are small companies with annual sales of less than $100 million. One result of their comparatively small size is that few possess adequate resources for addressing environmental issues (MCC 1993, p. 97).

The faster a company can erect a fab and start shipping chips from it, the faster it can recover the cost of the fab's construction. Paradoxically, it also may be the case that being first to market matters little if the firm is unable to produce sufficient numbers of chips. In 1997 Advanced Micro Devices (based in Sunnyvale, California) announced plans to release a microprocessor that would run nearly as fast as Intel's forthcoming Pentium II product yet would cost about 25 percent less. But even though AMD's entry may indeed erode some of Intel's market share, AMD lacks Intel's fabrication capacity. In 1996 Intel operated at least twice as many fabs as AMD, and by 1997 Intel was erecting a new fab roughly every 9 months (Kirkpatrick 1997, p. 63). The "fabless" Cyrix Corporation's inability to match Intel's volume is even more pronounced. Until 1997, Texas-based Cyrix sought to compete with Intel on the basis of design. Cyrix's strategy was to design a rival microchip and contract its manufacture to third-party suppliers, including IBM and SGS-Thomson. In 1997 Cyrix was purchased by National Semiconductor as part of its bid to build chips for network-based computing.

In addition to hitting economic limits, the quest to build smaller and faster microchips ultimately may collide with the laws of materials science. As channel widths continue to shrink, they become crowded with more and more electrons. The swarming electron hallways create increasingly strong electrical fields that may undermine a chip's performance. Experts remain divided as to when the limit on how many transistors can be squeezed onto a chip will be reached. However, most agree that this will occur sometime between 2006 and 2016 (Malone 1996, pp. 60–61). Beyond that point, it will probably be necessary to harness entirely new materials and manufacturing methods.

To help sustain advances in microchip manufacturing, US firms have jointly developed a "technology road map" (SIA 1994). In the short term, the industry may continue to achieve technological advances by using new materials. For example, IBM, Motorola, and Texas Instruments have developed methods that allow them to use copper instead of aluminum and thus to make faster, cheaper chips that consume less electricity and generate less heat. Further miniaturization may be achieved by switching from silicon to gallium arsenide or some other substrate material. Better photolithographic equipment capable of beaming smaller and smaller circuit patterns onto silicon is being developed. Other proposed chipmaking methods call for more exotic materials, such as proteins that use sunlight to act as photosensitive switching devices (Freemantle 1995).

Over the longer term, we may see the advent of organic molecules that self-assemble into semiconductors, or we may see costly photolithographic equipment replaced by microscopic "rubber stamps" (Bradley 1996; Cambou 1996; Robinson 1996). Even further down the road, photons may be harnessed to act as switches. When and if such radical technological shifts occur, today's state-of-the-art fabs will likely become industrial dinosaurs.

Defining Microchip Makers

As the foregoing discussion illustrates, it is often difficult to discuss the microchip industry without mentioning the Intel Corporation. Increasingly, however, Intel's business model is more the exception than the rule in the chip industry. Founded in 1968, Intel started out as a maker of memory chips. It introduced the world's first microprocessor, the Intel 4004, in 1971. Less than 3 years later, Intel released the 8008, a chip with 8 times the 4004's capacity. Though Intel eventually abandoned the market for memory chips to the Japanese, it maintains its command over the market for leading-edge microprocessors by rapidly releasing new products. Today, more than 80 percent of the world's personal computers are powered by Intel chips.

An important aspect of innovation is the ability to match a microchip with complementary products. However, rapid innovation does not guarantee market leadership. In order for its value to be realized, a microprocessor must have a product application. In 1979, IBM was seeking a 16-bit microprocessor to power its forthcoming PC. At the time, Intel and Phoenix-based Motorola were the industry's leaders. In 1980, to win the IBM deal, Intel launched Operation Crush, an effort to outdesign and outsell Motorola. Motorola never regained its market position. A young Harvard dropout, William Gates, helped to further ensure Intel's success when he structured DOS, the disk operating system for IBM's personal computers, to suit the capabilities of the 8088 microprocessor. Applications for Intel microprocessors expanded as IBM licensed the PC's design to Compaq and other makers of "clone" computers.

From this perspective, the microchip industry is not a free-standing; rather, it is part of a continuum with the hardware and software industries. Both computer design and software design typically are based on chip capacity. Furthermore, the release of new software and new hardware often is timed to coincide with the introduction of smaller, faster chips. Conversely, new software applications and new computing equipment help to fuel the demand for faster, more powerful chips. Windows 95 or Windows 98 will not run well on a 1986 microprocessor.

In addition to furthering continuous advances in software and hardware, chip advances occasionally spawn entirely new industries. For example, desktop publishing was born when Apple moved to the Motorola 68000 chip, which improved resolution and made possible the first inexpensive application of computer graphics.

One result of such integration is that policies aimed at chip firms will have consequences in the hardware and software industries. For example, an environmental policy designed to shorten the time it takes Intel to obtain an air permit may benefit Microsoft and IBM (as well as benefiting consumers eager to adopt the latest, fastest personal computer). Thus, when formulating technology policy or environmental policy, it is important to identify how the semiconductor, hardware, and software industries are related. The SIC codes fail to illustrate the degree to which these industries are connected.

Under current classification schemes, the electronics sector (comprising manufacturers of transformers, storage batteries, printed wiring boards, and home audio and video equipment) is designated as SIC 36. The semiconductor industry (SIC 3674) is considered a part of the electronics industry (SIC 36), as the first two digits of the classification code suggest. Computer and office equipment is listed separately in the machinery sector (SIC 35). Software firms have not been included in any SIC designation, because they have not been considered manufacturers. In 1970, there was not even an SIC code for the computer industry. Since 1994, however, a federal interagency task force and industry representatives have been rewriting the SIC codes. When complete, the new system will be called the North American Industry Classification System and also will cover industries in Mexico and Canada. In the interim, the SIC codes are the US government's primary method of tracking the economic trends and the toxic releases of semiconductor and other high-tech firms.

The Policy Context

Because of the growing prominence of their products, it is increasingly common for executives of the leading chip, software, and hardware firms to approach the federal government with demands. Chief among these demands are for regulators to adapt to the pace of technological innovation—a pace typically set by leading chip makers. In 1997, Intel's chairman and former CEO, Andrew Grove, told the *Washington Post:* "We operate on Internet time, which is about three times as fast as clock time. The government works on government time, which is about three times as slow. That's a nine-fold difference." (Corcoran 1997)

Although high-volume, high-profile microchip producers have found an audience in Washington, this handful of companies largely fails to reflect the changing composition of the industry. Consider that Intel's

Table 1.1
Firm size, by revenue, fiscal year 1996. Source: US Securities and Exchange Commission Edgar database 10K Reports, 1997. Available at http://www.sec.gov.

	Revenues	Primary products
Intel	$20,870,000,000	Microprocessors
National Semiconductor	$ 2,623,100,000	System-level products for fax machines, local- and wide-area networking, telecommunications
Advanced Micro Devices	$ 1,953,019,000	Microprocessors
Cypress Semiconductor	$ 528,400,000	Memory chips, programmable logic devices
Cyrix	$ 183,825,000	Microprocessors
SEEQ Technology	$ 31,338,000	LAN[a] chips

a. Local-area networking.

revenues for fiscal year 1996 came to $20.8 billion, while those of the fabless firm SEEQ Technology Inc. were around $31 million (table 1.1).[3]

One shortcoming of US Securities and Exchange Commission data is that they are developed only for public companies, and thus they exclude private firms. The US Department of Commerce develops aggregate measures of firm size at the county level for both public and private companies. When measured by total number of employees, the semiconductor firms operating in the United States are split almost evenly between large and small companies. The Department of Commerce's County Business Patterns reports listed approximately 940 firms under SIC 3674 as makers of semiconductors and related devices in 1994, the most recent year for which firm-size data are available. Of these, 504 employed fewer than 20 employees (figure 1.1). Some small firms manufacture discrete devices such as sensing equipment or develop prototype products for large firms, but a growing segment of startup companies design specialized chips (US DoC 1994). Though small firms have only 15–20 percent of large firms'

3. Information on revenues, investments, and litigation is readily available to the public through the Securities and Exchange Commission's Edgar database (US SEC 1997).

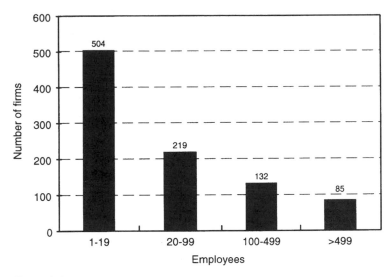

Figure 1.1
Firm size (number of employees), 1994. Source: US DoC 1996.

sales volume, they are an important source of innovation. Angel (1994, p. 94) observes that "many of these firms are on the leading edge of new product technology development and are growing rapidly." Because the Department of Commerce data deal with aggregates, however, it is impossible to determine how many individuals a single company employs. Public companies report their numbers of employees to the SEC. Not surprisingly, those numbers are consistent with revenues. For example, according to SEEQ's and Intel's annual SEC filings, SEEQ employed 74 in fiscal year 1996, while Intel's worldwide workforce approached 50,000.

According to County Business Patterns, about 85 firms employ more than 500. These very large firms most likely include "merchant" companies, including Intel, Motorola, and Texas Instruments, that design and sell microchips to makers of computers and other electronic devices. Such companies may employ up to 5000 at an individual facility. In contrast, a small chip firm may consist of one or two PhDs and 15 technicians.

Most of the other large companies are "captive" firms, such as IBM, Rockwell, and Hewlett-Packard, that design chips for use in the computers and other electronic products they design, manufacture, and sell. The categories "captive" and "merchant" are not mutually exclusive. For example, although IBM produces chips for in-house use, it recently began to fabricate microprocessors and other devices for other companies. A more exotic category of chip firms includes "fabless" semiconductor companies, which design chips and then contract their manufacturing out to suppliers. Some of the suppliers are captive companies (including IBM); others, known in industry parlance as "foundries," merely make chips and do not design or sell them. All these configurations are included under SIC code 3674. A company (e.g., IBM) that manufactures both semiconductors and computers may be assigned two four-digit SIC codes to reflect both activities.

Historically, semiconductor firms have clustered in California, Massachusetts,[4] Texas, Oregon, and Arizona (figure 1.2). Though most chip firms continue to maintain corporate headquarters in Silicon Valley, they are opting to build new manufacturing facilities elsewhere. New wafer fabs are often built in Texas or Arizona, each of which has an established base in chip and high-tech manufacturing. Unfortunately, it is not possible to use SIC data to track such trends, because the classification system does not differentiate among operational units within a company. Thus, no government data exist with which to determine the actual number of fabs in the United States. However, several industry trade associations have developed partial data sets. For example, Semiconductor Equipment and Materials International, a trade association for suppliers of fab equipment and materials, tracks firms' consumption of unfinished wafers by state, which may serve to indicate fab concentration. In 1994, Texas reportedly had the highest number of unfinished wafers, followed by Arizona and New Mexico (SEMI 1994). In addition to developing data on wafer consumption, SEMI identified at least 57 major fabs in existence in the American Southwest, and found that firms planned to build at least seven more in that region (table 1.2).

Similarly, no single public data source has been developed to keep track of the number of fabs in the United States and abroad. According to one trade organization, there were more than 900 fabs in 1993, 350 of them

4. Many of the Massachusetts firms classified as chip makers are entrenched Route 128 electronics firms.

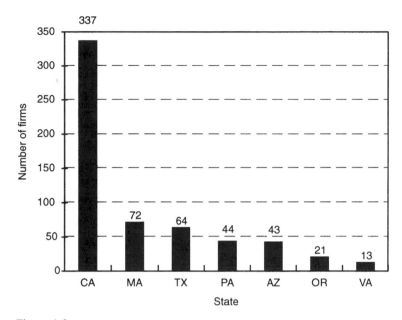

Figure 1.2
States with highest concentrations of semiconductor firms, 1994. Source: US DoC 1996. (The designation "firm" may or may not include manufacturing facilities.)

in the United States (MCC 1993, p. 97). Data on 70 international chip manufacturers show that in 1997 the world's fabs had the capacity to produce about 79 million 6-inch-equivalent wafers per year. If the average fab makes 30,000 wafers per month, and thus 360,000 per year, straight-forward division would indicate that the number of 6-inch-equivalent fabs currently operating is at least 219; however, such figures fail to indicate how many 8-inch and 12-inch fabs are in operation.

Some microchip firms are breaking with the tradition of geographic concentration to build fabs in regions that lack an established fabrication base. For example, two fabs were recently built in Virginia.

Foreign Competition

Until the late 1970s, US microchip manufacturers faced few foreign rivals. According to Morris (1990), about 85 percent of all breakthroughs in chip technology until that time were achieved by US manufacturers.

Table 1.2
Fabs of the Southwest, 1994. Source: SEMI 1994. (See table 1.3 below for a list of new fabs in the United States.)

			Number of fabs	Product
Texas	Dallas	Texas Instruments	3 + 1 new	Logic devices
	Arlington	National Semiconductor	2	Gate arrays, standard cells, local area network devices, advanced microcontrollers
	Carrollton	SGS-Thomson	3	Merchant and foundry producer
	Irving	Hitachi	1	DRAM[a]
	Dallas	Dallas Semiconductor	1	Memory devices
	Austin	Motorola	2 + 1 new	Logic devices
	Austin	Advanced Micro Devices (AMD)	3 + 1 new	Logic devices
	Round Rock	Cypress Semiconductors	1	Logic devices
	San Antonio	Sony Microelectronics	2 formerly owned by AMD	MOS[b] devices
	San Antonio	VLSI Technology	1	Gate array and PC chip sets
	Houston	Texas Instruments	3	MOS and bipolar devices
	Lubbock			
	Sherman			
	Corpus Christi	Sematech	1	Bipolar devices
Arizona	Phoenix	Motorola	12 + 1 new	MOS digital devices
	Chandler			
	Mesa			

Table 1.2
Continued

			Number of fabs	Product
	Chandler	Intel	1 + 1 new	Microprocessors
	Chandler	Microchip Technology	1	NA[c]
	Tempe	California Micro Devices	1	NA
	Tempe	Micro-rel	1	NA
	Tempe	Landsdale	NA; 1 new	Digital devices
	Scottsdale	Microsemi	1	Discrete devices
	Phoenix	SGS-Thomson	1	NA
	Tucson	Burr Brown	1	NA
New Mexico	Albuquerque	Allied Signal	1	NA
	Albuquerque	Intel	3 + 1 new	NA
	Albuquerque	Phillips	2	MOS devices
Colorado	Fort Collins	NCR (AT&T)	1	MOS memory devices
	Fort Collins	Hewlett-Packard	1	MOS logic devices
	Colorado Springs	Hewlett-Packard	1	MOS logic devices
	Colorado Springs	NCR (AT&T)	1	NA
	Colorado Springs	AMTEL	1	Logic and memory devices
	Broomfield	Microsemi	1	High power discrete devices
Utah	West Jordan	National Semiconductor	1	Logic products

a. Dynamic random-access memory.
b. Metal oxide semiconductor.
c. Not available.

However, US companies' focus on continuous innovation, in conjunction with the lack of foreign competition, masked serious shortcomings in US manufacturing methods. Angel (1994, p. 17) observes: "As long as US firms maintained a technological advantage over their international rivals, . . . production difficulties primarily affected profit margins, rather than market share." US firms still control the markets for microprocessors and custom chips, but Asian firms now control the market for memory chips.

The dominance of US firms started to slip around 1980, when they controlled about 60 percent of the world market. By 1996 that share had slipped to about 44.3 percent, and Japanese firms controlled about 36.7 percent (Dataquest 1997). Even Intel was not immune. During the mid 1980s, Intel, like most other US firms, largely abandoned its memory chip business to Japanese firms that were able to sell superior products at lower cost. During 1985 and 1986, Intel experienced cumulative losses of $250 million and laid off 6000 employees (Mazurek 1994).

Like Intel, most other major US microchip firms shed thousands of manufacturing jobs during the late 1980s. Advanced Micro Devices and National Semiconductor laid off 3000 manufacturing employees in Santa Clara County between 1987 and 1992 (ibid., p. 96). Today, most major US chip companies still maintain headquarters and planning and design functions in Silicon Valley, but chip manufacturing occurs in other Western states and, increasingly, overseas. Restructuring in the US industry may therefore be understood as an attempt to regain lost markets and to exploit new markets (Angel 1994, p. 20). In addition to streamlining the workforce, a common strategy for US firms has been to improve long-neglected manufacturing methods.

It is in the context of intensified global competition that the federal government in recent years has launched initiatives to promote the competitiveness of US chip firms. Recently announced public-private initiatives to fund chip research and earlier programs such as Sematech are among such efforts. In addition to industrial policies, two prominent Clinton administration environmental initiatives, the Common Sense Initiative and Project XL, also represent efforts to remove regulatory barriers to semiconductor, electronics, and computer firms.

Manufacturing Matters

It appears that federal efforts to promote the competitiveness of the US semiconductor industry may be paying off. For example, Sematech announced in 1997 that US firms would hire about 40,000 new technicians in the next 5 years to work in US fabs recently completed or under construction (Sematech 1997).

A fab technician typically earns $25,000–$38,000 a year, depending on the individual's experience and on the location of the fab. Once a high-volume fab is operational, the firm that owns it often tries to recover its enormous investment costs by running the fab 24 hours a day. To accommodate such a pace, fab technicians often work 12-hour shifts, four days on and three days off. Technicians in this historically non-unionized industry have frequent overtime opportunities. In addition to overtime opportunities, large chip makers often offer generous benefit packages and educational incentives to provide workers with sufficient technical skills.

From another perspective, some say, chip manufacturing jobs compare less favorably to serving coffee at Starbucks. In addition to long shifts and rarefied factory conditions, market swings prompt frequent layoffs and high employee turnover. Hossfeld (1995) has found that the low level of unionization also affects the composition of the industry's labor force. According to Hossfeld, the typical manufacturing employee in the semiconductor industry is "small, foreign, and female." By this account, chip and computer firms favor female employees because the industry assumes that women make more passive and obedient workers.

Fuzzy Logic, Fuzzy Boundaries

Though the federal SIC codes have always described the industry poorly, restructuring makes it increasingly difficult to identify precisely what constitutes "the US semiconductor industry." The Department of Commerce issues quarterly reports on the value of the products shipped by domestic manufacturers of semiconductors and related devices (SIC 3674). According to the Department of Commerce's *Current Industrial Reports*, in 1996 shipments of US-shipped semiconductors and related devices amounted to $67.53 billion (US DoC 1997b). ("Shipments" means the

Table 1.3
New fabs in the United States, 1994–1998. Sources: Company annual reports; SEMI 1994; Malone 1996.

		Firm	Completed	Product
Arizona	Chandler	Intel	1996	MPUs[a]
	Phoenix	Motorola	1996	Telecommunication devices
California	Santa Clara	Intel	1996	Advanced MPU design
	Roseville	NEC	1998	16- and 64-MB DRAM
Florida	Orlando	AT&T Microelectronics	1997	ASIC[b] graphics
Maine	South Portland	National Semiconductor	1996	Telecommunication equipment
New Mexico	Albuquerque	Intel	1995	MPUs
Oregon	Hillsboro	Intel	1996	MPUs
	Gresham	LSI Logic	1997	ASIC
	Hillsboro	Intel	1998	MPUs
	Gresham	Fujitsu	1996	16- and 64-MB DRAM
	Eugene	Hyundai	1997	16- and 64-MB DRAM
Texas	Arlington	National Semiconductor	1994	NA[c]
	Fort Worth	Intel	1999	MPUs
	Austin	Motorola	1995	SRAM[d] Power PC MPUs
	Austin	Samsung	1997	16- and 64-MB DRAM, ASIC
	Dallas	Texas Instruments	1996	Custom MPUs
	Austin	Advanced Micro Devices	1995	MPU Flash
Utah	Lehi	Micron Technology	1996	64-MB DRAM
Virginia	Manassas	IBM/Toshiba	1997	16-, 64-, and 256-MB DRAM

Table 1.3
Continued

		Firm	Completed	Product
	Goochland County	Motorola	1999	Telecommunication equipment
	Henrico County	Motorola/ Siemens	1998	64-MB DRAM until 2000, then SRAM
Washington	Puyallup	Matsushita	1997	4- and 16-MB DRAM, microcontroller units

a. Microprocessors.
b. Application-specific integrated circuit.
c. Not available.
d. Static random-access memory.

movement of products from the manufacturing plant; it does not necessarily reflect sales.) In 1996, the value of 16-bit microprocessor shipments by US firms was around $1.2 billion. Shipments of DRAM chips with capacities greater than 15 megabytes came to $1.4 billion.

Before restructuring, most companies that reported the value of semiconductor shipments were US-owned and integrated. Thus, the four-digit SIC code 3674 was sufficient to gauge the economic performance of the US chip firms. Today, however, a chip designed by a US company in Silicon Valley may be fabricated by a Japanese company operating in Texas or Taiwan and sold in Singapore. Yet the SIC data only reflect shipments of firms located in the United States.

Matters are blurred further by the fact that a growing number of Japanese, Korean, and European companies now operate wafer fabs and foundries in the United States. Table 1.3 lists all announcements of new fabs and fab expansions in the United States since 1994. Though most of these announcements came from US-owned companies, several came from foreign firms (including Fujitsu, Hyundai, and Samsung). Other new fabs are to be built by US-foreign partnerships. For example, Motorola and German-owned Siemens have announced a joint venture to build a fab in Virginia, as have IBM and Japanese-owned Toshiba. As table 1.3

Table 1.4
Shares of worldwide semiconductor market (estimated), 1996. Source: Dataquest 1997.

Rank	Company	Market share[a]
1	Intel	12.5
2	NEC	7.4
3	Motorola	5.7
4	Hitachi	5.7
5	Toshiba	5.7
6	Texas Instruments	5.0
7	Samsung	4.6
8	Fujitsu	3.1
9	Phillips	3.0
10	SGS-Thomson	2.9

a. Percentage of total revenues

shows, a number of existing fabs in the United States are owned by foreign companies, including SGS-Thomson, Sony Microelectronics, and Hitachi.

The Semiconductor Industry Association, an industry trade organization, uses data on 70 international chip manufacturers in developing its World Semiconductor Forecast. In contrast to the Department of Commerce's statistics, the SIA data reflect sales rather than product shipments and are global rather than national in scope. By the SIA measure, global semiconductor sales totaled $129.2 billion in 1996 and $153 billion in 1997 (SIA 1998). Table 1.4 lists the top ten global semiconductor companies by percentage of total sales (market share). Market share is one of the most convenient ways to categorize companies on a global scale. The data reflect worldwide sales of all semiconductors, including microprocessors, memories, and other products. In terms of market share for all semiconductor products, Intel's lead does not appear quite as commanding as its control of more than 80 percent of the microprocessor market.

Statistics that distinguish among product categories are illuminating because markets for different kinds of semiconductors tend to behave very differently. For example, the prices of DRAM chips dropped precipitously between 1995 and 1996, causing revenues to fall by 40 percent, and DRAM revenues fell another 14 percent in 1997. Conversely, microprocessor sales jumped by 30 percent in 1995, by 17.5 percent in 1996, and by 22.9 percent in 1997.

Even within the markets for specific products, prices, product release times, manufacturing methods, and volume tend to vary significantly. The variation is due to learning as well as to the spiraling cost of a new fab. For example, Intel typically dominates the market for the fastest new microprocessors both by releasing products ahead of AMD and Cyrix and by having the manufacturing capacity to outproduce its competitors. However, in the market for less powerful microprocessors Intel's competitors may give the giant a run for its money.

Most federal, state, and local policies to promote the competitiveness of US chip firms are based on less detailed information about market and firm structure. In the case of trade policy, it is likely that mathematical tools for developing realistic models simply do not yet exist. However, it is possible that those who make environmental policy simply assume that all firms are similar to Intel in products, size, and structure. In either case, failure to develop policies based on an accurate picture of firms and markets may carry unintended consequences. For example, a central goal of President Clinton's Project XL is the potential for facility-level regulatory experiments to be transferable to other firms. However, an experiment that reduces permitting delays for Intel is largely irrelevant to firms other than market leaders and is likely to be applicable only to fabs that produce leading-edge microprocessors in large volume. Though Intel may one day transfer the experiment to another Intel plant, it is questionable whether Intel's costly, contentious plan would make sense for a maker of memory chips, or even for one of Intel's two main rivals in microprocessor manufacturing.

States' strategies to lure wafer fabs must also learn to distinguish among chip markets. One example of why this is important is the fact that the sagging market for memory chips prompted Motorola to modify its plans for one of its new Virginia plants.

Summary

Rapid change in the semiconductor industry is driven primarily by the tendency of US merchant producers to pack more and more circuits onto a silicon chip. The rate of change has accelerated in recent years as competition has intensified and as the cost of new manufacturing facilities has skyrocketed. Unlike most products, chips are distinguished by the degree to which their value is based not on raw materials but on ideas. Traditionally, the clustering of firms in places such as Silicon Valley has promoted the exchange of ideas that drives continuous breakthroughs. Furthermore, since microchips are of little use without the products they drive, where the chip industry ends and the software and personal computer industries begin is difficult to distinguish.

Industry restructuring is blurring the once-clear understanding of how and where firms make microchips. In contrast to the popular perception, the category "semiconductor companies" includes both large integrated producers and smaller startup firms that may lack manufacturing capacity. Furthermore, the category "US industry" may no longer be useful. While the US government seeks to help a few prominent US companies remain competitive, a handful of Asian and European firms are—independently or in tandem with US producers—supplying new manufacturing jobs. Most new fabs are constructed outside Silicon Valley.

Despite the heterogeneity of chip firms, current policies to promote competition are largely tailored to fit Intel and a small number of other leading companies. Although being first to market may be advantageous to any of the leading firms, the spiraling costs of fab construction make it clear that a company is able to overwhelm its competitors simply by being one of the few firms able to bankroll a new fab. It is therefore possible that economic and environmental initiatives to help chip makers move products more rapidly to market may merely confer additional advantage on a few prominent producers.

2

Hitting a Moving Target

. . . as the position is better resolved, the momentum becomes more and more uncertain.
—Dickerson, Gray, and Darensbourg, *Chemical Principles* (1984), p. 306

The Heisenberg Uncertainty Principle illustrates the difficulty of assessing the environmental impact of microchip manufacturing: though it is possible to measure environmental performance at any point in time, estimating risks to humans and the environment associated with the industry's momentum is fraught with uncertainties.

This chapter reviews some of the best-documented environmental problems associated with chip manufacturing. By some measures, the environmental, health, and safety impacts of microchip manufacturing appear low relative to other industries. By other measures, risks to people who work inside wafer fabs may be among the highest in industrial production. Rapid innovation helps to explain the apparent contradiction. As the occupational health expert Joseph LaDou (1986, p. 2) observes, "the technology often moves so rapidly that new materials and processes replace old ones before sufficient information is obtained on the health effects of either." Though it confounds risk assessment, rapid innovation also creates unique opportunities for chip firms to improve product and process design faster and with less economic disruption than firms in industries where plants, processes, and equipment change less rapidly. In recent years a few prominent microchip firms have indeed made some significant strides in reducing—and in some cases eliminating—the use of a few problematic substances; however, such was not always the case.

Making Microchips

Strip away the plastic sheath that encases a computer. Short of electrical shock, it is hard to imagine how the grids of gleaming thumbnail-size squares "inside" (to paraphrase Intel's advertising slogan) pose a health problem. Indeed, the principal environmental issues associated with microchips stem not so much from how people use them in products, or even from how they dispose of individual microchips, as from how the chips are manufactured.

Microchips, unlike automobiles or even vacuum tubes, are resource intensive—when sold, they contain very few of the raw materials used to manufacture them. In theory, microchips command much less landfill space than automobiles or vacuum tubes. In practice, however, it is difficult to anticipate the degree to which chips may pose a waste challenge, because the products that chips power become obsolete long before the useful life of the chips is over (Parks 1997).

Though chips themselves may not yet pose much of a disposal problem, the substances used to ensure their purity represent both a resource-extraction challenge and a disposal challenge. One expert estimates that up to 99.9 percent of the materials used to manufacture chips are not contained in the final chip product (Glasser 1993). In terms of resource extraction, a thumbnail-size chip requires relatively large amounts of inputs, including chemicals, gases, and water. A 1993 industry-led study found that a fab producing 6-inch wafers used more than 2 million gallons of de-ionized water, 2.5 million cubic feet of high-purity nitrogen, and 240,000 kilowatt-hours of electrical power *per day* (MCC 1993, p. 97). Manufacturers are reluctant to report the actual amounts of materials used to fabricate chips for fear that competitors may appropriate the data. According to one popular account (SVTC 1995), successfully manufacturing one 6-inch wafer, which yields more than 200 chips, requires 20 pounds of chemicals and more than 3200 cubic feet of gases.

As the industry seeks to increase the number of transistors on a chip, the use of some inputs such as water and silicon may increase. As has already been mentioned, in order to increase the number of chips that can be placed on a single wafer, the industry is beginning to build fabs that make 12-inch wafers. Such a fab may use as much as 5 million gallons of water per day. Other environmental effects of packing more circuits onto a chip are less clear. For example, in recent years several leading manufacturers have developed methods that allow copper to be used instead of aluminum in circuits. Copper, a better conductor than aluminum, allows circuit lines to be made smaller. The substitution will increase performance, and it may reduce energy demand. The EPA considers copper and copper compounds as well as aluminum fumes and dust to be toxic. In certain forms, copper and its compounds are toxic in the environment. Aluminum fumes and dust also are known to have adverse effects. It remains unclear what the net environmental effects of a change from aluminum to copper might be.

Intensive use of chemical solvents and water helps to ensure that chip channels, which are some 200–400 times narrower than a human hair, remain free of microscopic contaminants, the leading source of device failure. To further guard against contamination, the most sensitive manufacturing processes take place in the sterile clean rooms, where workers clad in white head-to-toe "bunny suits" operate robotic equipment that ushers chips through a maze of stations containing clusters of costly production "tools." (Clean rooms take up roughly a tenth of the total area of a fab.) The protective clothing is designed not so much to shield the workers as to keep hair, skin, moisture, and germs away from the germinating chips. During a wafer's gestation, no human hand ever touches it. As a further precaution, all the air in the average Class 1 clean room is filtered roughly every 6 seconds. Purity standards allow no more than a single half-micron particle per cubic foot of air—roughly half a million times cleaner than the air in the average hospital operating room. To achieve such standards, a fab has about 10,000 tons of air conditioning equipment. More than half of the 240,000 kilowatt-hours of energy consumed per day is used to clean and condition the air inside clean rooms (MCC 1993, p. 97). In addition to the filtration, temperature, barometric pressure, and humidity are tightly controlled.

Fabulous Fabs

To understand the environmental problems historically associated with microchip manufacturing, it is useful to know something about how chips are made. In general, the production process involves the following phases: design, ingot growing, wafer manufacturing, cleaning, testing, and assembly (Van Zant 1990; US EPA 1995b, pp. 15–24). Manufacturing or fabrication, the most chemical-intensive phase of the process, typically occurs in clean rooms.

The number of manufacturing steps depends on the product and its intended use. Application-specific integrated circuits (ASICs) tend to require the fewest manufacturing steps. A memory chip requires roughly 100 steps. Microprocessors, considered the most complex chips to produce, may require more than 300 steps. The process is akin to baking a cake with eight to ten layers of tiny, patterned transistors. Some layers are embedded in micro-thin silicon; others are spread like icing on top. Today's chips are fashioned from polished silicon wafers about as thick as a credit card and from 3 to 8 (soon, 12) inches in diameter, depending on the function for which the chip is ultimately designed.

The first phase of the production process is product design. Computer models are used to develop and test layouts of circuit paths. The purpose of the design process is to create a master circuit pattern, known as a mask, which ultimately will guide the inscribing of circuitry on silicon. The most technologically sophisticated chips are designed years in advance of production. A number of large companies now develop manufacturing prototypes after the design stage to minimize production uncertainties and to improve wafer yield.

Wafer preparation is the first phase of manufacturing. At one time wafers were prepared in house, but now most firms purchase prepared silicon ingots from suppliers, most of which are in Japan. In raw form, silicon is a poor conductor. Therefore, ingot suppliers typically treat or "dope" areas of the silicon with ions to give it semiconducting properties. Once the ingots arrive at a fab, technicians slice and polish them into wafers. They then use a chemical etching process to smooth each wafer's surface and remove chemical impurities. Chemical etching employs cleaning solvents, acid solutions, and de-ionized water.

Fabrication is the heart of the manufacturing process and also the most chemical-intensive. Microchips are built, layer upon layer, by exposing some parts of a wafer to light and employing chemicals and water to etch away others. The wafer is then bombarded with ions to create semicon-

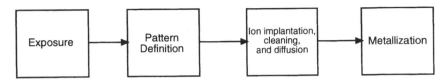

Figure 2.1
Wafer fabrication. Source: Van Zant 1990.

ducting regions. These regions are then wired together with a conducting metal, which may be just 0.35 micron wide. When complete, the wafer is rinsed, polished, and sliced into dies. The dies are tested, and defective ones are discarded. The next step, known as assembly,[1] involves attaching the metal strips ("leads") that allow the device to be connected to a machine.

In more detail, wafer fabrication is a repetitive process that consists of four basic phases: exposure, pattern definition, ion implantation, and metallization (figure 2.1).

Exposure requires using an oxidant, such as oxygen or steam, to create an oxide layer on the blank, polished surface of a wafer. The oxide layer protects the wafer during further processing. After oxidation, the wafer is thoroughly dried and cleaned. The wastes generated by oxidation include organic solvent vapors, rinsewater with organic solvents from cleaning, and spent solvents.

After oxidation, the wafer is treated with a brown, syrupy film known as photoresist. Negative photoresist blocks radiation from penetrating the wafer's surface; positive photoresist promotes exposure. The wafer is then inserted into a camera known as a "stepper," where light is beamed through the mask pattern created during the design process. The mask allows light to pass onto selected parts of the wafer. This process, which establishes the physical dimensions of the chip, is known as photolithography. Table 2.1 lists some of the chemicals that have been used in photolithography.

The next step, pattern definition, involves developing and removing the spent photoresist to reveal the oxide layer. After the photoresist has been removed, the wafer is put into a plasma that strips (in industry parlance,

1. Most US firms now send dies offshore for assembly and final packaging.

Table 2.1
Chemicals that have been used in photolithography. Source: US EPA 1995b.

Photoresists	Developers	Solvents and cleaning agents
Ortho-diazoketone	Sodium hydroxide	Deionized water
Polymethacrylate	Potassium hydroxide	Detergent
Polyfluoroalkylmethacrylate	Silicates	Isopropyl alcohol
Polyalkylaldehyde	Ethylene glycol	Acetone
Polymethylmethacrylate	Isopropyl alcohol	Ethanol
Poly(hexafluorobutylmethacrylate)	Phosphates	Ammonium hydroxide
Isoprene	Ethyl acetate	Xylene
Ethyl acrylate	Methyl isobutyl ketone	Ethylbenzene
Copolymer-ethylacrylate	Xylene	Chlorotoluene
	N-Butyl acetate	
	Cellosolve acetate	
	Isopropyl alcohol	
	Glycol ethers	

"etches") unwanted materials from the wafer's surface. Dry plasma etching forms a plasma above the surface to be etched by combining large amounts of energy with low-pressure gases. The gases usually contain halogens. Earlier "wet" etching techniques employed acid solutions. The dry plasma method provides better resolution of fine lines.

After circuit patterns are printed, electrically charged particles are implanted ("doped") into the silicon to form semiconducting regions. The step, known as ion implantation, creates a pathway for electricity to flow through the wafer. After ion implantation, the remaining photoresist is removed with chemical and plasma baths. The remaining particles are then diffused into the silicon wafer. Diffusion is a chemical process that exposes regions of the silicon surface to vapors of a metal additive (dopant) at high pressure.

Depending on the desired properties of the final chip, additional layers of silicon may be built up through chemical or physical vapor deposition. Chemical vapor deposition is a low-pressure process that combines appropriate gases in a reactant chamber at high temperatures to produce a uniform film thickness. Materials used during deposition may include silane, phosphine, diborane, nitrogen, and hydrogen.

After electrically active regions have been created on a chip, the components are connected ("wired"). The wiring consists of a thin layer of conducting metal, applied by chemical or vapor deposition in a process called metallization. Photolithography, one of the last steps in the sequence, fuses the electrically active components to the wafer's surface. Photolithography and etching also are sometimes used to remove any unnecessary metals. Wastes include acid fumes, organic solvent vapors, liquid organic waste, aqueous metals, and wastewater contaminated with spent cleaning solutions.

The four steps are repeated many times in order to build up layers of transistors and other circuitry. On average, it takes between 30 and 90 days for technicians and tools to gradually add and etch layers containing millions of transistors on a wafer. Over the course of chip building, certain manufacturing steps may be repeated up to 20 times, depending on the desired characteristics of the final chip.

Wastes generated in assembly include spent cleaning solutions, rinsewater with organic solvents, and excess thermoset plastic.

Environmental Effects

Manufacturers recently have eliminated the use of several prominent solvents in cases where the weight of scientific evidence shows potential risks to humans or the environment. They also have sought to minimize waste through increased recycling. Still, chip plants use, emit, and transport a host of constantly shifting substances that are known to be among the most toxic used in contemporary industrial production. To date, few studies have been developed to document the effects of substances on humans and the environment. Establishing causal links between chemicals used in chip manufacturing and human health problems such as cancer is complicated by the fact that the chemical mix is constantly changing. As a result, it is hard to determine precisely which chemicals may cause problems, and in what amount. Before reviewing suspected problems that are not well understood, it is instructive to review what is known about the environmental impacts of microchip manufacturing.

Valley of Superfund Sites

Chemical leaks from storage tanks to land and to groundwater have historically been among the most notorious problems associated with chip production. In October of 1984, the Santa Clara Valley Integrated Environmental Management Project identified 93 soil and groundwater contamination sites in Silicon Valley (Sherry 1985). Of the 93 sites, 63 were related to high-tech firms. Subsequent investigations by the US Environmental Protection Agency identified 29 sites so contaminated that they were added to the Superfund National Priority List of cleanup sites.

In addition to leading the United States in electronics technology, Silicon Valley carries the dubious distinction of having the highest concentration of Superfund sites. As of 1996, 20 of them were direct results of the production of wafers and other high-tech electronic components. Most of the contamination of groundwater is from leaking underground storage tanks. According to a 1995 report by the California Regional Water Quality Board, sites on the National Priorities List for cleanup include the following:

three Intel facilities in Mountain View and Santa Clara
three Advanced Micro Devices facilities in Sunnyvale
two Fairchild Semiconductor Facilities in San Jose and Mountain View
an IBM facility in South San Jose
two Hewlett-Packard facilities in Palo Alto
a National Semiconductor facility in Santa Clara
a Raytheon facility in Mountain View
a Siemens facility in Cupertino
a Teledyne Semiconductor facility in Mountain View
a Signetics (Phillips) facility in Sunnyvale
a Synertek facility in Santa Clara
a TRW facility in Sunnyvale
an Intersil (Siemens) facility in Cupertino
a CTS Printex facility in Mountain View.

The site that may be the most infamous is linked to Intel's first manufacturing facility at 365 East Middlefield Road in Mountain View. Located south of the San Francisco Bay wetlands, the site "has become one of the most problematic toxic plumes in the region," according to a report by Ted Smith, director of the Silicon Valley Toxics Coalition (Smith and Woodward 1992). In 1981, leaking solvents, including trichloroethylene (TCE), xylene, and 1,2 dichloroethylene (DCE), were discovered in groundwater. Some concentrations were up to 4000 times what health standards allow (ibid.).

Caused by a leaky underground storage tank, the toxic plume today is more than 6000 feet long and 500 feet deep. In the years since the site was discovered, the plume has mingled with similar toxic plumes from nearby Fairchild and Raytheon facilities and a nearby military base. The plume is slowly migrating underground toward the area's sensitive wetlands. Eventually it may enter the south portion of the San Francisco Bay. The EPA estimates that it could take up to 60 years to adequately clean up the contamination.

The groundwater contamination problem came into the public spotlight at Fairchild's facility in south San Jose. The contaminated site was publicized in January of 1982, when a number of residents in a south San Jose neighborhood began reporting health problems (Smith and Woodward 1992, p. 19). An epidemiological study by the California Health Department found three times the expected number of birth defects in the Los Paseos neighborhood near the Fairchild Semiconductor plant.

Fairchild, the progenitor of most major Silicon Valley chip firms, left another legacy. During the late 1970s, residents of San Jose started to experience high rates of miscarriages and birth defects. Although epidemiological analysis never directly established a causal link, some tied the reproductive abnormalities to Fairchild. The manufacturer's leaky fiberglass chemical storage tanks contaminated a major drinking-water well with high concentrations of 1,1,1-trichloroethane, a toxic organic cleaning solvent. The initial concentration of the solvent (5800 parts of solvent per billion parts of water) was 29 times California's recommended maximum level of 200 parts per billion. Long-term exposure to 1,1,1-TCA, also known as methyl chloroform, can damage nervous and cardiovascular systems. Exposure also can lead to dizziness, sleeplessness, and loss of coordination. Among the injuries claimed by residents in the neighborhood were congenital heart defects, childhood cancers, and thyroid, liver, and autoimmune disorders.

Harmful Gases

Groundwater pollution is the most visible and notorious problem associated with semiconductor and other high-tech computer and electronics companies. However, the industry's heavy use of solvents and gases also contributes to air pollution. Silicon Valley lacks the ubiquitous smokestacks, railyards, and scrap-metal heaps associated with Detroit, Pittsburgh, or Germany's Ruhr Valley. However, the heavy use of solvents in microelectronics production contributes both to the spread of large amounts of pollution across the country and to concentrated pollution with more hazardous chemicals, which are thought to contribute to more localized problems and which are toxic to humans and the environment in much smaller amounts than conventional pollutants.

Because they are typically released in amounts too small to trigger federal toxic reporting requirements, there is very little detailed information on the nature or the quantity of hazardous air pollutants (HAPs) released by wafer fabs. HAPs released in amounts large enough to trigger federal reporting requirements include hydrochloric acid, xylenes (mixed isomers), and methanol. Other substances released in smaller amounts include acetone, 1,1,1-trichloroethane, and methyl ethyl ketone.

Assessing the risks of HAPs is further complicated by the fact that they vary widely in toxicity and behavior once released into the environment. In some cases, 100,000 pounds of substance A may be much less hazardous than just 100 pounds of substance B. Upon release, some may persist in soil or in water; others may break down rapidly. In the same way that some substances are more toxic than others, chemicals take different exposure routes to affect human health. Some cause problems when inhaled; others are harmful only when they are ingested or when they come into contact with the skin.

Solvents used in chip manufacturing contribute to urban smog, too. Organic solvents vaporize as volatile organic compounds, which react with sunlight to form a major component of ground-level ozone. Anyone who drove through Silicon Valley on a bright summer day during its manufacturing heyday can attest to the region's air quality problems. Air pollution there was magnified by an onshore sea breeze that collected and pushed air pollution inland, where it was trapped against the mountains that circumscribe the valley. According to one study, the health risks associated with breathing the air in Silicon Valley during the 1980s outweighed the risks of drinking from the region's groundwater (Sherry 1985).

In addition to chemical solvents, microchip manufacturing calls for the use of hazardous gases to give wafers the desired electrical properties. The loss and rupture of a single cylinder of arsine from a delivery truck could deliver a lethal level of the gas for several square blocks and result in hundreds of fatalities. Other toxic gases used to make microchips include phosphine, diborane, and silane.

Streamlining Waste

Semiconductor firms in California have been targets of the state government's efforts to get them to comply with the Hazardous Waste Source Reduction and Management Act of 1989 (Senate Bill 14). The law is designed to improve what is known about firms' use of hazardous substances and hazardous materials. To promote source reduction, the state of California has identified the major hazardous waste streams generated by chip makers and the methods used by firms to reduce and control waste (California Department of Toxic Substances 1994, p. 12). Broader than "toxics," the definition of "hazardous waste" refers to a class of substances that may harm humans through *properties*, such as corrosivity, flammability, or infectiousness.

Sent to 300 semiconductor firms, the California survey identified hazardous wastewater as the largest waste stream. Most of the wastewater is generated through the manufacturing of ultra-pure, de-ionized water. De-ionized water is used primarily in the cleaning of silicon wafers. Wafers are fed through "wet benches," which consume roughly half of the ultra-pure, de-ionized water. To "manufacture" this water, firms treat incoming water to remove impurities. The process has efficiency rates around 50 percent. In other words, it takes two gallons of city water for a firm to manufacture one gallon of de-ionized, ultra-pure water.

Water that has been used in chip production often contains large amounts of acids (mostly sulfuric and hydrochloric) that must be treated before the water is returned to public supplies. A number of firms are

experimenting with costly and sophisticated systems such as reverse osmosis to remove acids; however, treated supplies may still carry high levels
of the sodium used to neutralize the acids, as well as small amounts of
acid.

Tracking Toxics

At the federal level, the most detailed data on the chip industry's toxic
patterns are contained in the highly imperfect Community Right to Know
requirements of the Title III provisions in the Superfund Reauthorization
Amendments (SARA). The Toxics Release Inventory (TRI) was created by
Congress in 1986 as part of SARA's Emergency Planning and Community
Right to Know Act (EPCRA). As part of EPCRA's provisions, states also
must develop State Emergency Response Commissions responsible for co-
ordinating certain emergency response activities and for appointing Local
Emergency Planning Committees to address chemical hazards. Legislators developed the emergency planning provisions in response to a 1994
chemical disaster in Bhopal, India.

The EPA devised the TRI under EPCRA's Section 313. The tool is intended to provide timely information to the public about firms' emissions
and transport activities. The TRI requires manufacturing firms that fall
within Standard Industrial Classification (SIC) Codes 20–39 to report on
emissions and transfers of 343 target chemicals. In 1994, the EPA increased to 643 the number of chemicals on which companies are required
to report. This increase, which took effect in 1995, makes it possible to
track emissions and transfers for more chemicals. However, the change
makes it difficult to directly compare reports for 1995 with reports from
previous years. In 1997, the EPA also expanded TRI reporting requirements to include seven additional industry sectors: metal mining, coal
mining, electric utilities, hazardous waste facilities, chemicals and allied
products distributors, petroleum distributors, and solvent recovery services (US EPA 1997b). The 1997 requirement calls for approximately
6100 new facilities in the seven industrial sectors to begin reporting on
toxic releases in local communities, bringing the number of facilities required to file annual TRI reports to 31,000.

The TRI exempts firms that employ fewer than ten workers or use
fewer than 10,000 pounds or manufacture fewer than 25,000 pounds of

TRI chemicals per year. Emissions and transfer data are based on estimates, rather than actual measures from the facilities that firms report to the EPA annually. The EPA compiles the data and supplies the information to the public in annual printed reports, computer tapes, on line, and on CD-ROM. As of 1997, TRI reports covered data from 1987 through 1995. The data from 1987 contain numerous errors and are therefore not used by researchers to assess industry or facility trends.

Despite its shortcomings, the Toxics Release Inventory has proved to be a powerful tool to promote awareness, among both industry and the public, of toxics release and transfer patterns. However, the TRI is a less useful tool for tracking toxics used and released in microchip manufacturing. The primary reason for this is that TRI reports include only a subset of the substances used to make microchips, because many of the substances are not listed or are used in amounts below the TRI's reporting thresholds. For example, in 1993 Intel used only 11 TRI chemicals in sufficient quality to require reporting. However, one fab may use more than 100 different chemicals. Also, because it is updated only once a year, the TRI fails to illustrate fluctuations in the nature and quantity of chemicals emitted.

Despite the TRI's shortcomings, industry, regulators, and interested citizens rely on it as a measure of facility, firm, and industry performance. The TRI is best used to indicate what amount of substances an individual facility uses in one year. However, a number of firms and environmental managers use TRI reports to assess performance over time. Here, the data are less useful to assess trends because the EPA since 1987 has added and dropped some chemicals on which firms are required to report, making it impossible to directly compare emissions over time.

With these caveats, the TRI shows that in 1995 semiconductor manufacturers released to the environment, or transferred for treatment and disposal, roughly 35 substances in quantities sufficient to trigger reporting. Some 139 facilities used SIC 3674 (semiconductors and related devices) as a primary identification code to file TRI reports in 1995. Total releases and transfers of TRI chemicals for these facilities came to about 20 million pounds (US EPA 1997b). To put such figures into perspective, consider that total TRI releases and transfers from all facilities that filed TRI reports that year exceeded 5 *billion* pounds.

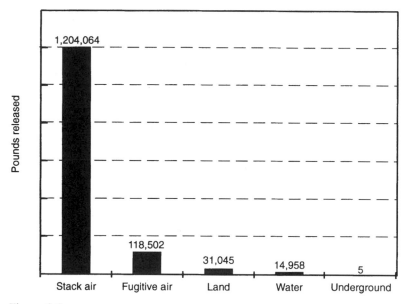

Figure 2.2
TRI releases, by medium, 1995 (totals based on facilities that use 3674 as primary SIC code). Source: US EPA 1997b. The computer search for facilities using this designation is based on a medium level of emissions detail. Searches for "high" or "low" levels of detail produce slightly different totals. Stack emissions: stack emissions "releases to air that occur through confined air streams (e.g., vents, ducts, and pipes)." Fugitive emissions: "those which are not confined to air streams (e.g., equipment leaks, evaporative losses from surface impoundments and spills, and releases from building ventilation systems)."

The TRI shows that off-site waste transfers far outstrip releases to the environment among chip facilities. Facilities released about 1.4 million pounds of TRI chemicals in 1995 and transferred another 19 million pounds off site for recycling, energy recovery, or treatment by a waste disposal company. Among toxics released into the environment by semiconductor facilities, the largest portion was released into the air. Facilities deliberately or accidentally released about 1.3 million pounds of toxics to air and about 15,000 pounds to water. Facilities released slightly more than 31,000 pounds of TRI chemicals to land (figure 2.2). The TRI chemicals most commonly released by facilities are sulfuric acid, hydrogen fluoride, and phosphoric acid. Others include xylene (mixed isomers), 1,1,1-trichloroethane, toluene, and methyl ethyl ketone.

In 1995, facilities that listed 3674 as a primary SIC code transferred equal amounts of waste off site to public sewage facilities and to energy recovery. Companies transferred 6 million pounds to publicly owned sewage treatment plants, and another 6.6 million pounds to energy recovery. Firms sent about 3 million pounds of TRI chemicals off site for recycling and another 2 million pounds off site for treatment. Reporting facilities disposed of more than 600,000 pounds.

When measured on the basis of weight, emissions and transfers of TRI chemicals from chip facilities are small relative to other industries. Consider that toxic releases and transfers from the chemical industry in 1995 accounted for almost 30 percent of all TRI releases and transfers, while releases and transfers from chip firms were less than 1 percent of the total reported by all firms to the TRI in the same year (US EPA 1997b).

In addition to the relatively small size of releases and transfers, TRI numbers for firms listed as SIC 3674 have fallen since 1987, when the EPA first required companies to file reports. Owing to data problems in 1987, reports from 1988 are a better basis for comparison. Between 1988 and 1994, releases and transfers from semiconductor facilities fell from a total of about 30 million pounds to 14 million pounds, roughly 53 percent. During that period, the number of facilities that listed SIC 3674 as a primary code fell from around 200 to 154. From 1994 to 1995, releases and transfers increased from 14 million pounds to slightly more than 20 million. It is likely that the increase in the number of TRI chemicals on which firms are required to report accounts for some of the change. In 1995 about 139 facilities filed TRI reports using SIC 3674 as a primary SIC code.

Before 1993, the TRI data provided few insights into why emissions levels fluctuate. Beginning that year, reports began to provide a method to help TRI users to gain insight into fluctuations and into the methods used by facilities to manage some types of on-site and off-site waste. The Pollution Prevention Act of 1990 requires firms that report to the TRI to also develop figures on TRI chemicals in a facility's waste stream. Firms also are required to report on methods used to reduce or eliminate waste. As a result, the TRI gives a fuller picture of how facilities manage some types of waste. For example, reports show that at Advanced Micro Devices' Fab 25 in Austin managers used a combination of source reduction measures, including improved maintenance and record keeping, and improved inventory control and storage methods to manage ammonia (US EPA 1997b). The TRI's production waste reporting provisions also require firms to show whether waste increases are proportional to changes in production levels (US EPA 1996g, p. 162). However, production indices are effective in adjusting TRI changes to output only in cases where firms emit the same chemicals from year to year—a highly unlikely scenario for leading-edge wafer fabs.

In practice, the public seldom has access to production information that could help people to assess why emissions levels change. As was noted in chapter 1, chip makers are reluctant to report on facility output and product identity for fear that such data may help competitors to climb more quickly the latest product learning curve.

Adjusting emissions for economic activity is somewhat easier when the data deal not with sensitive facility-level information but with industry-wide aggregates. From this perspective, the value of chip shipments for SIC 3674, in constant dollars, increased at annual rates ranging from roughly 4 percent to roughly 49 percent during the period 1988–1995 (US DoC 1997b). In other words: increases in shipments fluctuated widely from year to year, but the volume of chips shipped from factories increased over the entire period. Although such crude descriptive statistics still fail to identify the source of TRI declines, they help to rule out the possibility that drops in emissions and transfers are strictly due to falling wafer output.

Fuzzy Boundaries Revisited

The picture of environmental performance among semiconductor manufacturers also changes depending on the scale employed to analyze them. As was discussed in chapter 1, a case can be made for considering the chip industry as part of the broader computer and electronics sector. Though there are some exceptions, a chip is mostly useless without the products it powers. (Some exceptions include chips that can be tied to runners' shoelaces to measure race times and identification devices that can be implanted under the skin of costly livestock to prevent theft.) Emissions and transfers for the sector are reported to the TRI under the designations SIC 36 and parts of 35. In 1995, emissions and off-site toxic transfers from the electronics sector (SIC 36) came to roughly 447 million pounds (US EPA 1997b). To put the figures into perspective, consider that in 1995 releases and transfers from the chemical sector (SIC 28) came to around 1.7 billion pounds.

From a sectoral perspective, rapid change in the semiconductor industry also can be understood as driving rapid obsolescence of computers and other electronic products. Ted Smith, director of the Silicon Valley Toxics Coalition, put it more plainly (NSC 1997): "The business dynamic is to make everyone buy a new gizmo every 6 to 12 months." As chip manufacturers begin to overturn Moore's Law, computers and the products that run on them will become obsolete even more rapidly.

In addition to toxic releases and transfers, interdependence among prominent chip makers, computer manufacturers, and software firms is creating new waste challenges. Microprocessor manufacturing may call for up to 100 different types of chemicals and more than 300 manufacturing steps, but more than 700 compounds go into a computer workstation. Of the chemicals used to manufacture computers, about half are toxic according to the SVTC. One of the toxic chemicals contained in computers is lead, which manufacturers mix with glass in computer monitors to shield users from radiation. Mercury, cadmium, and phosphorus are trapped in the vacuum of the monitor's cathode ray tube. Inside, printed wiring boards hold small amounts of heavy metals. The batteries in laptop computers contain corrosive chemicals. Compact disks and floppy disks contain a metal that can contribute to acid rain.

The Environmental Protection Agency classifies personal computer monitors and circuit boards as hazardous waste. Disposal rules apply to corporations, but not to individuals who contemplate setting an old Apple IIe or IBM XT on the curb for the trash truck to conveniently cart away. Owing in part to the high toxic chemical content, obsolete computer parts are often difficult to recycle. Furthermore, most machines were not designed with disassembly in mind. Finally, few secondary markets exist in the United States for non-toxic computer parts that can be reused. Combined, such factors pose a mounting waste problem. And it is estimated that 10 million personal computers are discarded in the United States per year.

As manufacturers bring new chips to market more rapidly, the problem may get worse. A study by researchers at Carnegie Mellon University (Navin-Chandra 1991) estimates that for every three new computers manufactured at least two older computers become obsolete. As a result, nearly 2 million tons of computers will be contained in landfills by the turn of the century. The study further predicts that by 2005 the ratio of new to obsolescent machines may be nearly one to one. In other words, the rate at which personal computers become obsolescent is increasing.

Factors Complicating Risk Assessment

Though releases and transfers of TRI chemicals from chip firms appear small relative to those from other industries, the data are a poor measure of risks posed by chemicals to humans and the environment. A determination of risk depends on many factors, including whether or not exposure to the chemical occurred, the type of exposure, and a clear-cut identification of the mechanism through which a substance acts to trigger an adverse health effect. Comparing chemicals on the basis of weight or volume alone can create misleading impressions, because some chemicals are extremely toxic in small volumes or when used in combination with other chemicals. Indices of the toxicity to humans of the TRI chemicals show that the most harmful are more than a million times as toxic as the least harmful. For example, 10 pounds of phosgene may be more toxic under certain situations than 100 pounds of methanol (US EPA 1996g, p. 9).

Some researchers propose contrasting TRI data with rankings based on relative toxicity (Horvath et al. 1995). Weighting schemes could be used to better compare facilities, firms, and industries. However, such perspectives focus on controlling pollution rather than on preventing it.

The TRI data may contribute to misleading impressions of the risk from chemicals coming out of a facility. What the data say even less about is how the use of chemicals inside a facility may affect workers. Like most industries, semiconductor manufacturers must comply with employee health and safety standards set by the US Occupational Safety and Health Administration and enforced through reporting requirements and periodic or occasional inspections. Though the industry was not heavily regu-

lated during its infancy, equipment and methods to protect employees from chemical exposure are now well established (Williams et al. 1995). Indeed, by most accounts, the methods employed today by major chip companies to protect workers and the environment are state of the art. A number of these advances have been achieved through measures to reduce chip contamination as well as to reduce the time required to manufacture microchips. For example, the newest fabs employ high-efficiency particulate air filters. Filters remove contaminants and fumes from chemical work areas. To reduce potential delays, chemicals are increasingly ordered "just in time" rather than stored and piped directly into work areas. "Just in time" methods also may reduce the risk of accidents and spills. To develop and monitor such safeguards, most Fortune 500 chip companies also support staffs of environmental, health, and safety specialists, whose responsibility is to ensure that firms are in compliance with relevant laws and regulations (Davies and Mazurek 1996).

The methods employed by some very large companies in the chip industry today may be light years ahead of other manufacturing sectors, but some critics argue that was not always the case. A growing number of toxicologists and health experts suspect that the risks to humans of manufacturing microchips are just beginning to come to light.

A 1996 lawsuit filed by IBM fab workers in State Supreme Court in Manhattan claims that chemicals used to fabricate wafers caused serious illness among workers at IBM's facility in East Fishkill, New York (Glaberson and Campbell 1996). The original plaintiff, Keith Barrack, claims that his testicular cancer is related to exposure to chemicals used in IBM clean rooms. In an interview that appeared in the 25 May 1997 *San Jose Mercury News*, Barrack said that he was not made aware of the reproductive risks associated with some fab room chemicals until a friend showed him the warning label from a bottle of glycol ether, which said that the substance may cause testicular cancer. Barrack said he never thought to question the effects of chemicals used to manufacture microchips because he believed so strongly in his employer. Barrack is quoted as follows in the *Mercury News*: "IBM is a God in this area: my mother, grandmother and grandfather worked there: I turned cartwheels when I got on there. It was people before profits. IBM took care of its people." Now, Barrack blames IBM's chemical suppliers for his illness, saying

that the company known as Big Blue puts profits before people: "It's all profits first. It's push, push, push to get product out." (Hubner 1997, p. 26)

Since Barrack filed his lawsuit, more than 100 others have joined him. The IBM Case, as it is called, received national attention in 1997 when the NBC TV news magazine *Dateline* devoted a segment to the lawsuit. The plaintiffs allege that chemicals used in semiconductor production caused cancers. Eleven of the plaintiffs charge that the chemicals used to manufacture chips led to the death of their spouses. Another 13 women who worked at IBM claim that their employment caused cervical or uterine cancer. Another six men allege that chemicals caused testicular cancer, and another five claim that chemicals caused leukemia and other blood cancers. The remainder charge that chemicals caused everything from malignant brain tumors to birth defects.

The lawsuit seeks damages from four chemical companies that made compounds used by the computer and microchip manufacturing giant at East Fishkill. According to the *San Jose Mercury News*, the lawsuit targets chemical manufacturers because under worker's compensation law employees do not have the right to sue employers for injuries that are or may be work-related.

Barrack and the other plaintiffs in the lawsuit face an uphill battle. It is exceptionally difficult to link cancer to chemical exposure. Since 1970 the Centers for Disease Control (CDC) in Atlanta have studied more than 100 reported clusters of cancer in the United States that appeared to be linked to toxic chemical exposure. However, such studies show how difficult it is to link environmental factors to eventually development of cancer. Thus far, CDC researchers have been unable to establish a link at any of the reported clusters between chemicals and illness. The IBM lawsuit marks the first allegation that a cancer cluster is associated with fab room chemicals. Definitively proving that working in a wafer fab causes cancer is complicated by the industry's dynamism.

In 1992 Myron Harrison, a physician who had worked for IBM, wrote that the industry's quickening pace of technological change further hampers opportunities for plant health and safety professionals to identify and evaluate potential problems: "Any large semiconductor facility uses several thousand chemicals. An attempt to review the toxicology of all

these materials is doomed to be superficial and of little value." (Harrison 1992, p. 472)

In the case of carcinogens, there is usually a long latency—up to 30 years—between initial exposure to a substance and the actual development of cancer. However, demonstrating risk first requires effectively pinpointing the substances to which workers were exposed. Making such determinations in an industry characterized by high turnover in both process chemicals and technicians is extremely challenging.

A spokesman for the Semiconductor Industry Association, the leading trade group for the semiconductor firms, was quoted on the front page of the 28 March 1996 *New York Times* as follows: "The [SIA] is unaware of a single worker who has been exposed to a carcinogen and has a problem as a result of it." Indeed, even if evidence exists to establish that a worker was exposed to a suspected chemical or mixture of chemicals, much more data are required to prove that a chemical harms humans or other living things.

Determinations of cancer risk must be supported by controlled experiments where specific doses of a suspected carcinogen are administered to laboratory rats. Assessments customarily are reinforced by epidemiological studies, or reports which document health problems among people exposed to the substance in question. Indeed, under the Toxic Substances Control Act of 1976, the leading federal toxic law, the EPA does compel companies to initiate studies to identify potential carcinogens. The EPA can require manufacturers to initiate studies if risk or harm is suspected from animal data or chemical structure. The EPA also can require manufacturers to undertake health and safety studies if employees develop health problems. But no such studies have been developed on chip firms, in part because the IBM case represents the first time that technicians have publicly come forward to claim that their cancers were caused by working in a wafer fab. To date, the industry has not developed epidemiological studies or risk assessments to determine whether chemicals used in fabrication may cause cancer.

The East Fishkill case underscores just how little is known about the potential environmental problems associated with chemicals used to make microchips. According to Bruce Fowler, a University of Maryland toxicologist who tracks several chemicals used in chip production, further study will come about only when more illnesses are reported. When quizzed on the IBM Case, Fowler told the *New York Times*: "What I suspect is, over the next 5 to 10 years, we are going to start seeing more and more of these kinds of reports." (Glaberson and Campbell 1996)

The EPA requires more than health and safety data to control or ban use of toxic substances. EPA staff also must be able to show that the added economic cost to industry of tighter regulations is outweighed by risk reduction benefits. Consider the case of asbestos. Once commonly used as an insulation material and fire retarding material, exposure to some forms of asbestos have been linked to an increased incidence of lung cancer. When the EPA attempted to ban its use in pipe fittings, manufacturers took the agency to court. The Fifth Circuit Court vacated the EPA's proposed ban not because the agency had failed to prepare sufficient health and safety data, but because the EPA failed to adequately estimate the potential economic effects of banning the use of the substance in pipe fittings (*Corrosion Proof Fittings*, 947 F.2d 1201, Fifth Circuit Court 1991). Making an economic case for eliminating substances used in chip manufacturing may be even more challenging due to the lack of potential substitutes to ensure similar levels of chip cleanliness, as the following case of glycol ethers illustrates.

Reproductive Problems

To date, the most closely studied health issue among semiconductor workers involves shorter-term human health effects linked to a group of solvents commonly used to apply and strip spent photoresist. Glycol ethers are a long-time staple of semiconductor manufacturing. However, the industry did not begin to document their potential health effects until 1986. Subsequently, three independent studies on thousands of women who worked in wafer fabrication facilities found that subjects had higher than average rates of reproductive problems. A 1988 Semiconductor Industry Association study conducted by researchers at the University of California at Davis found miscarriage rates among fab workers to be 40 percent higher than the rate of women in the industry who work outside wafer fab rooms. The California study also found that women who worked in fabrication areas required more time to become pregnant than those who did not work in fabrication areas. A 1992 study conducted by researchers at Johns Hopkins University on IBM employees found that women who work in fab clean rooms had an increased risk of spontaneous abortion as compared to women working in other clean-room processes that did not require glycol either use. The studies were preceded by a smaller research effort conducted in 1986 by the Digital Equipment Corporation. Though the Digital study never established the causal agent, it found miscarriage rates among fab workers to be double the norm of women working in other parts of the plant.

According to one account, industry studies predated the reports linking glycol ethers to reproductive problems by 15 years. Furthermore, evidence linking the solvents to other toxic effects preceded the studies by nearly 50 years. Despite the existence of such data, firms continued to employ the solvents through the early 1990s (Froines et al. 1995). As an interim solution following publication of the Digital study, some manufacturers simply discouraged women from working in wafer fabs. However, a number of firms have since introduced viable substitutes to glycol ethers and many firms have established voluntary goals to completely eliminate ethylene-based glycol ethers from the manufacturing process.

As early as 1980, before the industry glycol ether studies, some statistics showed that the industry was safer than other sectors. Semiconductor firms had just 7.6 injuries and illnesses per 100 employees, compared with 13.1 per 100 for all manufacturing industries. Semiconductor manufacturing does indeed minimize some workplace hazards associated with traditional heavy industry, such as injuries sustained form machinery with dangerous moving parts. However, the relatively low occupational hazard rates also are colored by the fact that occupational disease is grossly underreported for all industries and also by the composition of the semiconductor industry workforce. The semiconductor industry has a higher proportion of non-manufacturing employees than such heavy industries as automobiles, petrochemicals, and mining. LaDou adjusted the hazard rates to account for workforce composition and found occupational illness rates among semiconductor maintenance workers in California to be 3 times those of manufacturing workers in general (1.3 versus 0.4 per 100 workers). Most of the semiconductor occupational illness cases involve systemic poisoning amplified by high concentrations of chemicals in rarefied "clean-room" environments.

Public and Regulatory Response

In contrast to older industries such as automobiles or chemical manufacturing, the US semiconductor industry evolved around the same time as federal laws designed by Congress to control pollution to air, water, and land, as well as protect occupational health and safety. Like most industrial pollution sources, semiconductor manufacturers must comply with this set of individual laws, as well as local ordinances. For example, firms must obtain permits that specify how much air and water pollution the plant emits. Plant personnel also must complete detailed manifests to track hazardous waste. As illustrated by the glycol ether case, firms also must report health problems and file annual estimates of toxic releases and transfers. As a further precaution, chip makers must follow health and safety laws that establish standards for chemical exposure, handling and work methods.

Though most of the laws contain reporting provisions to help indicate whether the laws are controlling pollution, very little useful pollution data exist to show whether the laws are helping to control pollution (Davies and Mazurek 1997). As an illustration, consider that approximately 70,000 chemicals are listed under the Toxic Substances Control Act (TSCA), passed by Congress in 1976 (15 USC 2605–2609). However, by 1990, the EPA had compiled complete test data for only six chemicals (US GAO 1990, p. 3). TSCA's Section 4 establishes an Interagency Testing Committee (ITC) to recommend to the EPA which substances should be tested and which tests are necessary. By 1990, the ITC designated 386 chemicals for testing (ibid.). Insufficient information on the potential effects to humans and the environment of chemicals used in chip manufacturing makes it difficult for regulators to both assess and make an effective case for tighter regulatory controls.

Similarly, the Occupational Safety and Health Administration (OSHA) has set Permissible Exposure Limits for less than 1 percent of the thousands of chemical products in commercial use (Hawes, p. 4). Moreover, industry critics maintain that occupational health and safety standards have simply failed to keep pace with the rapidly changing industry. Silicon Valley attorney Amanda Hawes told the *San Jose Mercury News* that OSHA standards "are the equivalent of setting the speed limit on Santa

Clara Street at 500 miles per hour" (25 May 1997). Critics also maintain that health and safety agencies simply lack the resources to adequately police industry. For example, the *Poughkeepsie Journal* reported in 1996 that no OSHA inspectors had been inside the IBM East Fishkill facility in at least 10 years (Hubner 1997, p. 28).

Shifting Paradigms

In recent years, a growing number of observers have started to advance the simple idea that, rather than control pollution with Byzantine reporting requirements and costly equipment, laws and policies should encourage firms to design products and devise production processes that prevent harmful substances and materials from being used in the first place (Gottlieb 1995).

Known variously as *pollution prevention, toxics use reduction,* and *clean production,* the emerging approaches are primarily distinguished by the degree to which they strive to reduce the use of chemicals that may pose risks to humans and the environment. Pollution prevention tends to be the broadest, allowing for a hierarchy of management strategies, including recycling and reuse. In contrast, toxics use reduction and clean production seek to curb or eliminate toxic and hazardous chemicals.

Prevention-based approaches aim to help product planners, health and safety experts, and plant engineers focus on the entire production process, from resource extraction, to use, recycling, and disposal, rather than on using costly technology as waste disposal methods to capture and treat pollution (Freeman et al. 1992).

To promote practices that avert rather than control risks, Congress in 1990 passed the Pollution Prevention Act. The act encourages firms and other sources to voluntarily adopt methods to prevent or reduce, rather than control, pollution, but contains no enforceable provisions. Instead, the law and ensuing initiatives encourage companies to adopt a "hierarchy" of strategies to manage pollution. Methods to reduce the inputs that wind up as waste are at the top of the pyramid, followed by reuse, recycling and disposal. In response to the statute, the EPA launched a number of voluntary initiatives to promote prevention. Among educators and prevention advocates, Design for the Environment (DfE) is one of the most popular EPA initiatives because it requires industry and environmental managers to consider the impacts of manufacturing during product development, rather than after pollution is generated (Allenby and Fullerton 1992, pp. 51–61).

DfE helps product planners and designers create goods that economize on the use of materials which could otherwise end up as pollution. Though DfE is among the most popular methods endorsed by many environmental educators and advocates, such proactive programs must ultimately compete with a company's primary mission of meeting the bottom line. Some studies have demonstrated that prevention techniques pay for themselves by economizing both on materials and treatment costs (Freeman et al. 1992, p. 637). But it is far from clear that prevention is cost effective in all cases. Even when it can be demonstrated that prevention is more cost effective than control, product and process redesign efforts are often eclipsed by other company priorities (Greer and van Löben Sels 1997).

Regional Institutions

In addition to prevention-based approaches, a number of local ordinances and interest group initiatives in traditional chipmaking enclaves help to fill gaps in federal laws. For example, municipalities in Silicon Valley have developed at least three ordinances that deal specifically with microchip manufacturing. One ordinance is designed to prevent the recurrence of toxic chemical spills such as the Fairchild leak. Another requires strict monitoring of how firms use and transport gases such as arsine. The third is a county ordinance to purchase only from companies that have developed goals to rapidly phase out the use of chlorofluorocarbons, which are linked to global warming and depletion of the earth's protective upper-level ozone.

Although some ordinances appear to impose more burdens on chip makers, air quality officials in some localities also have developed regulations that allow semiconductor manufacturers to make necessary process adjustments without notifying authorities each time such a change occurs. For example, the Bay Area Air Quality Management District establishes different record keeping and reporting requirements tailored to the specific nature of semiconductor firms. Similar local permitting provisions have been developed near Phoenix, as well as for one of Intel's facilities in Oregon (Hatcher 1994).

Several public interest groups also have helped to promote worker and public understanding of risks associated with semiconductor and high-tech manufacturing. The SouthWest Organizing Project (SWOP 1995, pp. 22–28) and similar groups take the position that the lack of strong unions, coupled with the composition of the industry's labor force, may help to explain why health problems remain underreported. Based in Albuquerque, the SWOP has responded to an incident at a now-closed facility owned by General Telephone and Electric. Approximately 225 former workers, largely Latina, have been involved in lawsuits against GTE stemming from suspected cancers and other illnesses which they maintain stemmed from working at the plant (ibid., p. 23).

One group that has been at the front lines of the campaign to make the industry more accountable is the Silicon Valley Toxics Coalition, founded in 1982 in the wake of the Fairchild incident and similar discoveries associated with the high concentration of chip firms. The SVTC has helped to push environmental problems associated with microchip manufacturing to the top of local and federal policy agendas. The group successfully developed the nation's first community ordinance that mandates safer containment and monitoring of hazardous chemicals. The SVTC also is an active participant in current federal efforts such as the Common Sense Initiative, designed to improve how regulations target the computer and electronics industry. More recently, similar organizations have sprouted up in chipmaking enclaves in Texas, Arizona, Oregon, and Massachusetts.

The Industry's Response

To the degree that the industry's dynamism complicates efforts to track chemicals and assess their risks, rapid change presents chip companies with a tremendous opportunity to plan and design for continuous environmental improvements. Indeed, when challenged with strong scientific evidence and the threat of stiffened regulation, semiconductor and other high-tech firms have, in some cases, been able to phase out problematic substances more rapidly and with less economic disruption than some other industries with lower rates of capital turnover.

In historically stable industries such as chemical manufacturing, processes and equipment may last up to 15 years and plants much longer (Kirschner 1995). For such industries, modifying processes or machinery to improve environmental performance can require significant modification and costly delays. Because the equipment and process methods inside a state-of-the-art wafer fab are routinely modified to achieve continuous chip refinements, environmental improvements are not necessarily as disruptive.

The CFC Case

The case of chlorofluorocarbons (CFCs) best illustrates how the chip industry's dynamism can be compatible with cleaner production. By the 1980s, the weight of scientific evidence identified CFCs as a major contributor to the depletion of the stratospheric ozone level. The solvents were widely used by electronics, computer, and chip producers to clean sensitive parts, such as disk drives and the printed wiring boards to which chips and other electrical components are attached. Though they pose little or no health risk at ground level, their properties in the upper atmosphere made CFCs a target for phase-out under the provisions of the 1987 Montreal Protocol. Because of the importance of CFCs' solvent properties to computer and electronics producers, the industries were initially opposed to the requirements of the treaty. At the time of its signing, electronics and computer companies claimed that they lacked viable substitutes for CFCs.

However, when participants adopted the Montreal Protocol, the position of the computer and electronics industries quickly changed. Maxwell et al. (1993) document how the electronics and computer industry had mostly eliminated the use of CFCs by 1992.

As an illustration of how rapidly such goals were attained by individual corporations, consider that Intel in 1990 set as a goal the elimination of ozone-depleting chemicals from manufacturing processes by the end of

1992. Intel achieved a 98 percent reduction by 1992 and eliminated the remaining 2 percent by 1994 (Intel 1995, p. 15).

Why were electronics, computer, and chip companies able to achieve reductions so rapidly? According to Maxwell et al. (1993), the relatively short product cycles that characterize computer and electronics manufacturing helped to make it easier for companies to comply with the treaty provisions than industries where capital equipment and chemicals processes are modified less often.

AT&T Microelectronics' Allentown works, a Pennsylvania site for the manufacture of sophisticated silicon semiconductors, provides a close-up of how one facility eliminated CFCs. In 1993 the facility had eliminated CFC emissions, down from a height of more than 300,000 pounds in 1982. The phase-out was achieved by implementing improved housekeeping and conservation strategies and through the development of alternative materials. Environmental health and safety specialists identified areas of potential waste by creating chemical use and emissions inventory (Fraust et al. 1992).

The Clean Air Act Amendments of 1990

Like the companies in the CFC case, Intel initially expressed concern with stepped-up efforts to control air pollution in the 1990 amendments to the Clean Air Act (Sheppard 1995). Congress designed the 1990 provisions to help states achieve cleaner air quality by requiring potential new emissions sources to revise permits when production changes that effect emissions occur. Permits must spell out all Clean Air Act Amendment requirements that apply to the air pollution sources, including scrubbers, boilers, and vents, as well as spell out compliance schedules.

Intel representatives testified before Congress that stepped-up permitting requirements could interfere with continuous improvements necessary to successfully manufacture chips. They maintained that permitting requirements could require Intel to undergo lengthy permit review and revisions each time it wanted to make a change in a process involving chemicals that contribute to air pollution. In a hearing before a congressional subcommittee on the proposed amendments, Intel representatives testified that lengthy regulatory review and permit revisions could cause the firm to "lose [its] top position in the market place" (Sheppard 1995, p. 5).

Much as in the Montreal Protocol experience, Intel made a remarkable turnaround after Congress adopted the 1990 amendments. The company responded to the new legislation by adopting a corporate policy to structure all of its new US facilities as minor sources for conventional and hazardous pollutants. One exception is an older facility in Oregon; another is a possible expansion in New Mexico. In general, minor sources release less than 100 tons per year of each of the six conventional pollutants and roughly 25 tons per year of pollutants listed under the 1990 Clean Air Act Amendments as hazardous. Designing mammoth manufacturing facilities to qualify as "minor" air emissions sources represents a major environmental achievement for Intel (PNPPC 1995).

Intel's ability to design all its new US facilities to fall within minor source designations was partly achieved by aggressive efforts to reduce emissions from solvents used to ensure chip cleanliness and clean manufacturing equipment. Many solvents used in chip manufacturing result in volatile organic compounds (VOCs), a major component in the chemical and physical atmospheric reactions that form ozone and react with sunlight to form urban smog.

Intel's facility in Aloha, Oregon, houses some relatively old production lines and is considered a major source of VOCs. There, Intel worked with state regulators to develop a novel pollution prevention permit under the 1990 amendments to the Clean Air Act. The permit approves in advance routine production changes which could otherwise trigger lengthy review or approval. The permit also approves in advance construction of an additional manufacturing facility. In exchange for flexibility, the permit conditions require Intel to describe what types of processes and decision making the facility will use to achieve pollution prevention reductions. The permit requires Intel to develop a data collection system and employee training program to promote pollution prevention. Finally, the permit requires Intel to establish partnership agreements with its suppliers and equipment vendors to minimize both hazardous and volatile organic compounds from raw materials (ODEQ 1994, pp. 11–12).

"Advanced Planning"

A number of companies have sought to improve competitiveness by opting to no longer manufacture microchips. But those that continue to make chips in high volume have instituted organizational changes to compress the time between product planning and sale. To reduce potential production delays, a number of large firms now develop manufacturing prototypes years in advance of actual production. This increased focus on product planning and design, called "advanced planning" in the jargon of the industry, concomitantly shifts the environmental focus away from the factory floor and toward equipment procurement and factory design.

Historically, semiconductor manufacturers addressed equipment safety during installation or in the course of actual manufacturing. Equipment safety codes were varied—a jumble of existing codes, standards, and company policies. In 1988, as chip manufacturers began to restructure, Intel sought to develop more uniform environmental, health, and safety guidelines (EHS) for manufacturing equipment performance (Intel 1995, p. 8). The purpose of the move was to integrate EHS concerns into initial equipment design and create long-term goals and objectives. In 1991 the Semiconductor Equipment and Materials International (SEMI), a trade organization, published product safety guidelines that mirrored those developed by Intel. In 1993, with broad input from the industry, SEMI revised and released a set of final safety guidelines for manufacturing equipment (SEMI 1993). Since then, a number of semiconductor companies based in the United States use the SEMI guidelines as a minimum EHS requirement for equipment purchases. The SEMI guidelines encourage chip and equipment manufacturers to focus on EHS issues in the design phase. They also encourage producers to obtain environmental, health, and safety assessments by independent third parties. Equipment suppliers in the United States and abroad have adopted some variant of the guidelines into product safety programs.

Greater emphasis on production planning is one reason behind Intel's successful strategy of minimizing air pollution from its new US fabs. Intel's production and fab construction increased in the early 1990s partly because of successful restructuring efforts and a rising demand for personal computers. Between 1990 and 1994, Intel's semiconductor production increased 98 percent. But during the same period, the company reports, VOC emissions increased only 18 percent (Intel 1995, p. 15). In terms of VOC emissions per square inch of silicon processed, emissions fell by 75 percent over four successive process generations (Intel 1996a, p. 11). It is important to note that Intel used a mix of strategies, including improvements to manufacturing equipment, to achieve the lower rate of increased VOC emissions. Other strategies include improved chemical utilization, solvent substitution, and better abatement methods (Intel 1995, p. 15).

One of Intel's strategies had unintended negative repercussions. Solvent substitution caused inadvertent odor problems at four fabrication facilities. To help reduce VOC emissions, Intel substituted ethyl 3-ethoxypropionate (EEP) for a more volatile solvent (Intel 1995, p. 15). At Intel's largest facility, located in Rio Rancho, New Mexico, the EEP odors prompted complaints from neighboring residents and contributed to poor community relations. To correct the problem, Intel in 1994 spent $16 million on thermal oxidizing units at all four facilities to destroy the odors and reduce VOC air emissions (ibid., p. 15).

The Intel case helps to demonstrate the degree to which semiconductor companies are capable of rapidly reducing the use, release, and transfer of some substances. Indeed, owing to such successes, a number of companies increasingly maintain that industry is more qualified than government to improve how companies manage pollution (Davies and Mazurek 1996). The SEMI guidelines illustrate how environmental concerns can be compatible with economic goals, such as shortening time to market. Though industry-led efforts to improve product and process design must be applauded, some question whether companies would make such moves absent today's federal laws.

Critics concede that the industry is much cleaner today than it was 20 years ago, but most agree that improvements were mostly driven by federal environment, health, and safety laws. One such critic is Gary Adams, a former IBM chemist. In the aforementioned story in the 25 May 1997 *San Jose Mercury News,* Adams described how work conditions changed dramatically during the 1970s, after Congress expanded federal pollution control laws. When Adams first went to work in IBM's San Jose plastics lab, during the mid 1960s, he and his co-workers "had [their] hands" in solvents that would "go right through rubber gloves" (Hubner 1997, p. 9). Years later, Adams developed a malignant tumor on his left femur. Adams eventually developed a list of ten former colleagues who had worked in or near an IBM plastics lab in San Jose from the 1960s to the mid 1970s. Among those listed, eight developed cancer and seven have since died (ibid., p. 12).

Fred Tarmann, one of Adams's former co-workers in the lab, told the *Mercury News* that after the federal laws were expanded, IBM began to remove chemicals that he and his colleagues had formerly used on a daily basis. The company imposed tighter controls on chemicals and developed calibration methods for fume hoods to ensure that work areas were properly ventilated (ibid., p. 11). Tarmann credits federal laws with the changes: "Companies won't do it for themselves because it detracts from the bottom line." (ibid.) The occupational health specialist Joseph LaDou similarly told the *Mercury News*: "I have no doubt that without the federal government, companies would be handling chemicals today the way they did 30 years ago." (ibid.)

Federal pollution control laws are largely based on conventional economic theories about why firms pollute: in the absence of regulation or properly defined property rights, releasing pollution to air, to water, and to land is free, or what economists call "external" to the firm's manufacturing costs. Regulation makes the cost of pollution apparent to the firm and serves to limit the amount of pollution that a firm seeks to release.

Conventional theory and wisdom suggest that laws, standards, and regulations are the most effective mechanisms to ensure that environmental goals are met. Unfortunately, economists have found that traditional methods are not necessarily the most efficient. For years, economists have maintained that market-based instruments such as emissions trades could get the job done more inexpensively. Environmental and justice groups prefer other strategies because they claim that market-based instruments fail to consider how risk may be unevenly distributed across populations. The Silicon Valley Toxics Coalition, for example, has historically relied on information-based strategies and occasional public protests to compel companies there to improve environmental, health, and safety performance. The SVTC also has successfully promoted the adoption of more stringent local health and safety ordinances. For example, Santa Clara

County passed a toxic gas law in 1990, 3 years after the SVTC issued a public report on the potential hazards associated with arsine and other toxic gases used by the industry.

The toxic gas ordinance has since become a national model. Intel was the first major manufacturer to substitute liquid arsine for arsine gas at all its facilities, thus reducing the threat of a catastrophic incident. The arsine replacement was achieved through Intel's "Operation Leapfrog," which evaluates potential replacement chemistries to reduce impacts on the environment and improve worker safety (Intel 1995, p. 8).

As the arsine case suggests, the threat or adoption of stricter regulations and information campaigns are often complementary. In the case of hazardous chemicals, the threat of liability as a result of Superfund, coupled with adverse publicity surrounding the industry's 20 contaminated soil and groundwater sites in Silicon Valley, has prompted major firms to improve chemical storage methods. For example, in 1985, shortly after contamination from several of its leaking underground storage tanks became apparent, Intel voluntarily removed all its underground storage tanks worldwide and replaced them with double-reinforced, aboveground storage facilities. Today, most major chip firms have replaced the earlier generation of tanks with double containment tanks. And, whereas companies once put tanks and pipes underground and out of view, today they are above ground and often equipped with monitors to promote leak detection.

In other cases, individual companies have improved management as a result of direct public opposition. For example, a group of Silicon Valley protesters marched on IBM in 1989 after learning from the data that CFC releases from one of the company's Silicon Valley facilities were the highest in the nation. The group demanded and received from the manufacturer a pledge to reduce and phase out use of the substances (US EPA 1990, p. 323). Similarly, in 1991 the Campaign for Responsible Technology persuaded Sematech to initiate an environmental research and development program (SVTC 1997, p. 7). Other information-based approaches target more select audiences; for example, one public interest group used the stock market in efforts to promote environmental accountability at Intel.

Occasionally, technological innovation simultaneously eliminates some environmental problems. One example involves manufacturers' efforts to improve chip resolution. To achieve smaller geometries the industry in recent years has shifted from wet chemical etching to dry plasma methods. Dry etching processes include various halogenated or nonhalogenated gaseous compounds, as well as radio frequency power. Wet etchants are usually acids such as sulfuric and phosphoric, as well as substances such as ethylene glycol or solutions of ammonium, ferric, or potassium compounds. Though the switch reduced use and disposal issues associated with wet solvents, the potential health effects of dry methods are less well understood.

Systems currently under development to further promote smaller chip geometries also may help to further reduce the potential for human exposure to toxic chemicals. Sematech's National Technology Roadmap for Semiconductors includes a plan to help companies meet environmental health and safety challenges as the industry shifts to smaller chip geometries (SIA 1994, pp. 66–180). Goals include the reduction of hazardous chemicals use, energy, and water use reduction and stronger programs to protect factory workers (ibid., p. 80).

The push to make smaller, better chips is helping to reduce some potential routes of semiconductor worker exposure to chemicals. For example, the industry is promoting greater use of robotics, as well as vacuum encasements or "microenvironments," to protect wafers while they are inside tools and during transport around the fab. Though primarily designed to reduce wafer contamination, such advances also may reduce worker exposure. To appreciate the degree to which wafer fabrication now takes place in a sealed environment, consider that technicians in some areas of the newest fabs are no longer required to wear bunny suits. Innovation, coupled with deliberate industry efforts to reduce occupational health and safety problems, may reduce the odds that today's fab room worker will be a plaintiff in a cancer-related lawsuit 20 or 30 years from now.

While automation is reducing some exposure routes to humans, the impact of innovation on resource use and the environment is less clear. Consider the industry's 1997 shift from 8-inch to 12-inch wafers. The shift to larger wafers increases the number of potential chips each wafer yields. Intuitively, such shifts obviously will require more of some inputs such as silicon. Some say that the industry change-over also requires up to 2 million more gallons of water per day than a 6-inch-wafer fab, but estimates remain unsupported.

Over the longer term, the potential incentive to firms to make environmental, human health, and safety gains may be offset by the abbreviated life of current manufacturing methods. While experts remain divided on when firms will finally reach a limit to the number of transistors that may be squeezed onto a chip, most agree that this will happen between 2006 and 2016 (Malone 1996, pp. 60–61). After that, it is likely that firms will need to harness entirely new materials and manufacturing methods to create semiconductors. Given such a short time horizon, what incentive do chip firms have to ensure that today's greenfield fabs are not tomorrow's contaminated brownfield sites?

In terms of soil and groundwater contamination, CERCLA's cleanup and liability provisions have obviously prompted chip firms to adopt sounder chemical storage methods. However, Congress has been under consistent pressure from a number of industries to relax some of Superfund's provisions, and the future of the statute's strict liability features is by no means certain.

Prevention Barriers and Opportunities

Because risks associated with technological innovation are so difficult to predict, a growing number of observers advocate methods to prevent problems before they are created. Prevention-based approaches seek to shift the focus of environmental management from its current focus on control and treatment toward product planning and design strategies that reduce or eliminate pollution. Such a perspective incorporates environmental concerns all the way up the supply chain to input choices. To conserve on materials, for example, analysts must understand why certain substances are used to manufacture products, as well as be able to identify viable substitutes. Another method may require assessing product planning and design decisions to evaluate where hazardous materials might be removed from the final product (Gottlieb 1995).

The SEMI equipment guidelines illustrate how the industry's planning focus is compatible with environmental concerns. But barriers to prevention-based approaches persist. A 1993 industry study led by the Microelectronics and Computer Technology Corporation found that pollution prevention concepts had largely not yet found currency among most US semiconductor manufacturers. The study found that process development and design disciplines do not consider the environmental impacts of steps integral to operation (MCC 1993, p. 110). In other words, rather than integrate environmental considerations into design and manufacturing, most firms continue to favor control rather than prevention methods.

The MCC study identified significant prevention opportunities for firms to economize on the amount of materials used and waste generated. Indeed, the average semiconductor facility used resources such as water, energy, chemicals, and equipment inefficiently. "There is," the study noted, "a general attitude that in order to maintain the required level of cleanliness, the more chemical or water rinses used, the better." (ibid., p. 104) In this regard, the MCC study is consistent with the findings of the economic geographer David Angel (1994), who argues that the focus of US firms on continuous innovation masked manufacturing weaknesses.

The MCC findings were developed through an in-depth analysis of specific processes and chemicals at a wafer fab that at the time represented the industry norm. For example, researchers focused on photoresist, a light-sensitive chemical used to define circuit patterns. Photoresist is one of the most expensive chemicals used in wafer fabs and also carries one of the highest disposal costs. MCC researchers found the actual amounts of monthly chemical purchases far exceeded actual amounts required in the process. They found that the disparity was due both to poorly designed processes and inventory controls. The substance was applied in excessive amounts but also purchased in excessive amounts. Because photoresist has a short shelf life, unused supplies were routinely discarded (MCC 1993, p. 105). Through better inventory control methods, the firm was able to discard less photoresist and save 40 percent of the total purchase cost of the material, which translated to about $1 million for the fab studied.

The MCC study also found that wasteful practices were reinforced by poor monitoring and accounting systems. For example, researchers found energy and materials use were best measured through continuous, real-time measurement devices on each manufacturing tool, rather than estimation techniques. Similarly, the study found significant chemical use reduction in companies that have implemented cost accounting systems that "allocate true costs back to the users and generators." As of 1993, most fabs still maintained accounting systems where environmental costs were not directly considered as fab operation costs (MCC 1993, p. 109).

As the MCC study reveals, some of the largest prevention opportunities in high technology appear to be improved "low-tech" housekeeping methods. Examples include better inventory control, and practices to extend the life of some process materials. For example, chip firms in the California waste minimization study reduced aqueous waste streams primarily by extending the life of acid bath solutions and employing better bath filtration methods.

The MCC report posits that as global chip competition continues to intensify, manufacturers may find that making cleaner microchips can serve to reduce manufacturing costs. In time, companies may also use environmental performance as a way to differentiate their products. In addition to speed and performance characteristics, consumers may seek to purchase "green" chips or investors purchase stock in companies with distinguished records on the environment. Already, most large public chip companies, including IBM, AMD, Texas Instruments, Intel, and National Semiconductor, issue statements on environmental performance in their stock reports. Most major chip companies also release annual environmental health and safety reports. Reports are published and, in most cases, posted electronically on company web sites.

Prevention Opportunities Downstream

Since computers obviously contain more materials than the chips that run them, it may be that some of the greatest opportunities for environmental improvements are not in chips but in the electronic and computer equipment powered by them. The EPA has sponsored projects with Digital, Hewlett-Packard, IBM, Apple, and Intel to design more energy-efficient computers and to create workstations that can be more easily dismantled for reuse and recycling (MCC 1993).

One area where computer manufacturers have made progress is in manufacturing equipment that is increasingly energy efficient. The Computer Business Equipment Manufacturer's Association (CBEMA) noted in a June 1992 press release that computer power consumption had decreased by four orders of magnitude since 1952. Introduction of the semiconductor, which consumes a fraction of the electricity required of the vacuum tube, accounts for some of the gains.

Though computers are more energy efficient, the proliferation of desktop machines and the amount of time users spend on them are driving up energy demand. According to the MCC study, during the life of an average personal computer, the customer ownership phase uses the largest amount of energy (MCC 1993, p. 270). The EPA estimates that computer systems by 2000 could account for about 10 percent of total commercial electricity. The designation, commercial electricity, represents about half of the total electrical load (US EPA 1992).

To reduce energy demand, the EPA worked with the CBEMA to create the Energy Star program for desktop workstations and computer printers. The feature potentially reduces power wasted during periods when the machine is on but is not in use. The Energy Star feature reduces energy use to 30 watts or less when the machine is not in use. The feature has the potential to reduce 90 percent of power that would otherwise be wasted (MCC 1993, p. 270).

It may be that the environmental impacts of chips and computers are eventually offset by even larger rearrangements of human activity enabled by the emerging technologies. Several studies have suggested that after industrialization, societies begin to "dematerialize" (Herman et al. 1989, pp. 50–69). In other words, as economies mature, they move away from heavy industry to less resource-intensive forms of economic activity such as information technology or services. It is true that the microchip contains fewer material inputs and demands less electricity than its unwieldy predecessor, the vacuum tube. But improvements in energy and resource efficiency may be offset by industry use of water and chemicals necessary to prevent chip defects.

It may be that chips and the machines they power will one day help to reduce environmental impacts in other sectors of the economy, such as paper and fuel. Eventually, information technology may enable widespread work and communication methods such as telecommuting, teleconferencing, and "paperless" office communication. At present, the proliferation of computer and electronics equipment appears to be fueling opposite patterns in paper consumption (Smith 1997, pp. 45–54). According to the American Forest and Paper Association, use of computer printer and photocopying paper has increased by 30 percent since 1987. According to paper producers, the trend is being fueled by the growing amount of information generated by people using computers and the Internet (O'Harrow 1997). The biggest increase (in percentage and pounds) in paper use occurred in printing and writing paper—primarily office paper. In 1996, per capita consumption of such paper in the United States was 213 pounds, double the 1966 amount.

Semiconductors and the products they power and support also have the potential to reduce environmental impacts in other sectors, such as transportation and construction. Telecommuting and teleconferencing could cut air and water pollution associated with automobile, bus, and air transport. Such shifts also could help reduce the need for new road and office construction. Chips, furthermore, may contribute to greater fuel and energy efficiency by enabling the development of smarter manufacturing equipment and consumer appliances. But gains may be offset by consumers' greater reliance on electronic and computer gadgetry.

Akin to the way in which the internal combustion engine "solved" the problem of ubiquitous urban horse manure, the microchip likely will solve some old problems as it creates new ones. Though chips may someday help minimize auto-related air and water pollution, the proliferation of chip powered machines in the home may create new hazards to human health. For example, printers and fax machines emit solvents that can harm indoor air quality. Increased use of computer and electronic devices in the home also may increase the incidence of electrical shock or fires.

Clearly, forecasting the future net environmental impact of sweeping shifts enabled by the microchip is challenging. Today such exercises remain beyond the scope of even the most sophisticated models. As Oliviero Bernardini and Riccardo Galli (1993) observe, the difficulty of forecasting of energy and raw materials requirements remains a considerable obstacle in strategic policy decisions.

One of the chief challenges associated with forecasting is that it requires a global framework. A global framework helps to balance the impact of technological innovation in developed countries with the highly uncertain technological trajectories of many nations in the developing world.

As the next two chapters show, the scale issue is relevant to measuring the environmental performance of US semiconductor firms. For example, whereas 200 facilities reported to the Toxics Release Inventory in 1988, the number dropped to 139 by 1995. It may be that 61 facilities improved environmental performance so significantly that they no longer triggered TRI reporting thresholds or that companies employed chemicals not listed among the TRI's 643 chemicals. Chapter 3 tracks the performance and structural changes of five firms to show why the former scenario is highly unlikely. One reason that fewer facilities appear on the TRI is because a number of companies, spurred by the skyrocketing cost of building and equipping new wafer fabs, have contracted manufacturing out to firms located largely in developing nations. Chapter 4 shows how relocation of wafer fabs from Silicon Valley to other parts of the world further complicates efforts to use national measures such as the TRI as an adequate measure of the industry's environmental performance.

Summary

Contrary to their clean image, US semiconductor firms are a significant source of groundwater and soil contamination, as well as of certain types of air emissions and water effluents. Compared to emissions from some other industries, however, those from chip fabs appear minuscule.

However, when chips are understood as a fundamental part of the computer and electronics sector, the industry's total emissions rank only behind those from the chemical and metals sectors. Furthermore, the pace of chip production fuels a mounting waste pile of obsolete computers and electronic equipment.

The TRI shows that emissions from firms required to report also have decreased substantially since 1988, the first year for which comprehensive data are available. Data from several facilities show that firms achieved such reductions largely by improving recycling and reuse, rather than by changing designs to economize on materials.

Emissions data alone may understate the impact of the industry on humans and the environment because the data are a poor proxy for risk. Several staples of semiconductor manufacturing have been linked to reproductive problems, as well as to deterioration of the earth's ozone layer. Some suspect that chemicals used in semiconductor manufacturing will eventually be linked to other health problems, including cancer and birth defects. However, efforts to assess chemicals are complicated by rapid process change.

To the degree that it complicates assessment, rapid change also carries an enormous potential to prevent risks before they are generated in the first place. Semiconductor, computer, and electronics firms have demonstrated the ability to rapidly phase out chemicals such as CFCs or glycol ethers when the weight of scientific evidence demonstrates risk and stronger regulation therefore appears imminent.

Although better environmental management has certainly contributed to the declines of a few substances such as CFCs and glycol ethers, dramatic changes in both how and where firms produce chips also contribute to the appearance of improved environmental performance.

3

Real Men Have Fabs

American semiconductor companies are changing both how and where they manufacture microchips. Whereas design, fabrication, and assembly once took place under a single factory roof, a chip engineered and sold by a Texas firm may now be manufactured by a competitor in Phoenix, New York, or Taiwan, assembled in Singapore, and sold in Los Angeles.

This chapter examines the economic factors that have helped to reshape how and where firms manufacture microchips. The argument advanced here is that restructuring is an effort by US firms to improve their long-overlooked weaknesses in production methods (Angel 1994). As long as US firms faced few foreign competitors, there was little need to focus on product quality and manufacturing methods. US firms instead largely pursued a strategy of continuous technological innovation. But increased international competition, coupled with the complexities and growing cost of producing smaller and smaller chip circuits, have prompted some firms to pool costly wafer fabs and others to abandon manufacturing entirely. Firms that continue to construct wafer fabs increasingly locate them outside Silicon Valley, in some cases outside the United States. Though there is no evidence to suggest that firms locate new fabs in regions where environmental regulations are weak, it is true that some of the localities most anxious to lure wafer fabs possess the fewest public resources to protect human health and the environment. This chapter examines in depth three manifestations of restructuring: "fabless" firms, strategic partnerships, and "free-standing" fabs erected far from companies' research and design functions. Each case illustrates how economic and geographic changes complicate efforts to evaluate and manage the industry's environmental performance.

Roots of Restructuring

Restructuring is largely a result of three mutually reinforcing factors: maturing technology, greater international competition, and an innovation trajectory that is trained on producing ever-shrinking microscopic circuitry. Each of these factors is, in varying degree, related to the industry's unique innovative characteristics.

Maturing technology gives firms the flexibility to construct fabs anywhere in the world. Traditionally, uncertainties involved in chipmaking meant that manufacturers kept fabs near research and design facilities. As US firms began to focus on methods to improve long-neglected production methods, many reorganized to reduce uncertainties inherent in wafer fabrication. One outgrowth of corporate reorganization is that manufacturers no longer need to keep fabs close to corporate headquarters.

Maturing technology also increased competitive pressures by greatly expanding the number of transistors worldwide. For a time, US companies maintained unbridled industry hegemony because they generated and controlled the technical details about chip design and manufacturing. However, as the experience of the Traitorous Eight shows, chip manufacturing and design methods are hard to contain. As the industry matured, knowledge spread, enabling more manufacturers to enter the market.

Another factor motivating industry shifts stems more directly from US manufacturers' focus on achieving continuous technical advances. Making smaller and smaller transistors requires increasingly expensive manufacturing equipment. About 75 percent of the cost of a new wafer fab is equipment cost. As the investment required to design and produce chips continues to rise, firms have an even greater incentive to move products more rapidly to market to recover investment costs. Shorter product cycles further discourage investment in costly state-of-the-art wafer fabs.

Silicon Valley Fever

A recent TV commercial for Intel gave viewers a glimpse "inside" a stylized Intel wafer fab in order to tout a new chip with multimedia power. In the spot, fab technicians clad in colorful bunny suits make microprocessors to the disco tune of "Stayin' Alive," the 1970s anthem from the film *Saturday Night Fever.*

Indeed, as chip companies announce plans to construct costly megafabs in the United States and elsewhere, governments around the world have contracted "Silicon Valley Fever." Today even the most unlikely territories are willing to gamble development budgets on the chance of bagging a billion-dollar chip plant. Consider the state of Delaware. Fed historically by heavy manufacturing and more recently by financial services, economic development directors there have sought to enter the chipmaking development game by launching a high-profile effort to attract semiconductor manufacturers. Other countries also have entered into the frenzy. Egypt recently established a "Pyramid Technology Park." Malaysia's prime minister reserved 468 square miles for the purpose of growing a "Silicon Valley" (*Economist* 1997, p. 6).

Luring new wafer fabs is a high-stakes venture. Most observers agree that lowering the cost of traditional inputs ("factors of production"), including land, labor, and capital, can help make a location more appealing. However, that tactic can lead development officials to erroneously conclude that lowering land costs and regulatory requirements are sufficient to attract semiconductor companies. William T. Archey, president of the American Electronics Association, was quoted in the 30 January 1997 *Washington Post* as having said that the key factor in luring chip firms is an "appealing" economic and regulatory climate.

Indeed, as the construction cost of a new fab approaches $3 billion, companies are anxious to offset capital costs with state incentive packages, including tax breaks and worker incentives. States then submit competing proposals to the prospective company for consideration. In 1994 the *San Jose Mercury News* obtained a copy of California's proposal for a new Intel manufacturing facility known as "Fab 11." Reporter Rebecca Smith (1994) disclosed the details of the package in a piece that illustrates how the state "selection game" is played: "California is pitted ruthlessly against other states in blind bidding, eager to show it is competing aggressively for the kind of jobs that once flowed automatically to the Golden State." According to Smith, states submitted proposals to Intel patterned after Intel's "Ideal Incentive" matrix—a list of 104 requests, including a free utility service and roadways, as well as enticements such as a 50 percent price break on employees' moving expenses. Among the states that received the Intel list were Oregon, Arizona, Texas, Utah, and New Mexico.

Intel's Ideal Incentives

Table 3.1 lists some of Intel's requests. As part of its bid, the state of California pledged to waive approximately $8.1 million in fees associated with obtaining expedited permits. The state also offered Intel discounts on hotel and car rental rates. California's total package of $40 million was ultimately not sufficient to lure the fab. Intel representatives said that the decision to build the fab in New Mexico was motivated in part by a California tax on manufacturing equipment that would have added more than $70 million to the cost of constructing the plant.

Table 3.1
Source: Smith 1994.

Intel's request	California's response
Low or zero income tax	9.3 percent
Increase air emissions by 100 tons per year	No
No protection of flora and fauna	Minor review; no cost to Intel
Guaranteed 3-month permit timetable	6½-month guaranteed fees and staff paid
Off-site roads to meet needs with no cost to Intel	Free interchange
Air fare, hotels, car rental 10–20% below corporate rate	Yes
Employee moving and storage 50–60% below standard rate	$1000 when new home purchased
Employee title fees discounted at 1–2% of home purchase price	$700 discount on mortgage cost
Apartment rental discounts of 5–10%; 50% off deposits	No
In-state tuition status for employees and dependents	Prohibited by state statutes

New Mexico agreed to finance debt stemming from Intel's Rio Rancho plant with 30-year bonds. Furthermore, New Mexico eliminated property taxes for the manufacturer by agreeing to purchase the land where the fab would be located and leasing it back to Intel. Another factor in Intel's decision to build Fab 11 in New Mexico (and later, Fab 12 in Arizona) was the presence of older Intel facilities in Rio Rancho, New Mexico, and in Chandler, Arizona.

Ultimately, a central consideration for Intel in locational decisions is the degree to which states can condense the construction and permitting approval process. One Intel official told the *San Jose Mercury News*: "Time is money. If it takes us longer to get up and running on Site A than Site B, that can cost us billions in lost revenue." (Smith 1994) Such state offers are common. For example, the Texas Department of Commerce helped to lure the chip maker Samsung to Austin by offering incentives similar to those offered to Intel in New Mexico. The Samsung deal included industrial revenue bonds, sales and use tax exemptions, enterprise zones and $2000 per employee to train the 1000 new hires Samsung needed for its Austin facility. The state also promised Samsung streamlined permits and unspecified "regulatory assistance" (Texas Department of Commerce 1996).

Although companies typically seek to locate in a state with an existing base in chip manufacturing, Virginia recently secured three new fabs by holding out a generous package of state incentives. Virginia successfully lured three new wafer fabs (two of which have been built) with a total of $153.7 million in tax credits, grants, and worker training to develop a high-tech manufacturing industry, according to an article in the 16 June 1996 *Washington Post*.

Despite such offers, many regions anxious to replicate Silicon Valley have found that financial and regulatory incentives are not sufficient to lure chip fabs. For example, though Virginia has successfully lured two fabs, and a third is planned, it still trails far behind Texas, Arizona, and Oregon in high-tech job growth, according to a survey by the American Electronics Association. Indeed, AEA president William Archey said that financial incentives, although important, are strengthened when a state can show that it also supports a strong base of high-tech firms. A number of academic observers generally agree that it takes more than tax and regulatory incentives to lure knowledge-based industries.

Economic geographers have found that certain industries cluster in specific places—for example, the entertainment industry in Los Angeles and clockmakers in Switzerland's Jura Arc. For some industries, proximity offers both direct and indirect benefits. High concentrations of similar types of production can lower the cost of labor and provide a ready pool of trained professional and technical talent. The same applies to suppliers and specialty service industries, such as electronic and chemical manufacturers and hazardous waste removal specialists. Industry "agglomerations" also spread the costs of public roads that ferry supplies and schools that train future workers (Scott and Angel 1987).

Some say that proximity has other, less obvious but equally important benefits. Putting people with the same ideas and information together in a place like Silicon Valley can increase the probability that engineers will achieve more technical breakthroughs by increasing the odds that they will bump into one another in the workplace or in less formal settings such as coffee shops (Storper and Walker 1989). Indeed, economic development officials in Delaware admit that their chance of luring a chip plant is "a fairly long shot" because the state lacks an established base in semiconductor and high-tech electronics manufacturing (Behr 1997).

Just as technological breakthroughs and external economies helped to develop Silicon Valley, diffusion of product learning and of manufacturing know-how can cause clusters to break apart (Storper and Walker 1989, p. 61). As product knowledge improves, firms are free to locate production in the places that offer the greatest return, either because of low labor cost, because of proximity to foreign markets, or because of amenities such as clean air and affordable housing. Chip manufacturing in the United States tends to be highly localized, with the highest concentrations of companies in the desert Southwest. (See tables 1.2 and 1.3.)

In addition to creating predictable geographic patterns, maturing design and chip making methods also increase the number of potential foreign competitors. Although the knowledge-intensive nature of chip manufacturing makes it among the most secretive of industries, technical information is nevertheless difficult to contain, as the Traitorous Eight first demonstrated. Today, at least 37 firms trace their ancestry to Fairchild. Fairchild's offspring, and other companies such as Texas Instruments and Motorola, were responsible for the majority of technological breakthroughs in chip manufacturing during the industry's early days. But information regarding chip making methods is difficult to contain, particularly in an industry that thrives on entrepreneurs like the Traitorous Eight. Engineers and scientists outside the United States gradually learned to emulate, and in the case of Japan improve on, American manufacturing methods.

Somewhat paradoxically, a few leading US firms responded to the increasing competition by making it possible to build new fabs far away from research and design shops. Although chipmaking remains a highly uncertain enterprise, US manufacturers have gradually learned to reduce some of the uncertainties by simulating manufacturing conditions years before a fab is built. Furthermore, Intel and AT&T have assembled design teams from around the globe to simultaneously create new products. By reducing technological uncertainties surrounding production, it is possible for firms to plant state-of-the art fabrication facilities almost anywhere (Mazurek 1994; Angel 1994).

To some extent, the migration of new fabs away from California may be seen as a continuation of organizational changes that started in the 1970s. During the industry's tender years, most phases of production—research, design, manufacturing, assembly, and marketing—were performed by one company in a single location. By the early 1990s, such configurations had changed dramatically. By then, the majority of Silicon Valley companies either operated design laboratories and production prototypes or had dispensed with manufacturing entirely to focus primarily on customized product design.

According to a 1987 study, two-thirds of new semiconductor firms lacked their own wafer fabrication facilities, contracting such functions out to foundries instead (Rappaport 1992, p. 22). Another survey found that by 1991 only 31 of Silicon Valley's firms continued to operate fabs there (Gordon and Kreiger 1994).

Today, a few large-scale producers continue to operate fabs in Silicon Valley. The facilities do not produce products for mass markets, but instead serve as laboratories where the latest technological breakthroughs are first put through experimental production methods years in advance of constructing fabs for their manufacture worldwide (Angel 1994, p. 15).

Leaky Technology

Restructuring is a result of developments that occurred in the 1970s, when the Japanese first sought to emulate US success in very large-scale (multiple-transistor) chip technology. The Japanese government sponsored large-scale chip research efforts which seized and improved upon existing US memory chip production methods (Forester 1993).

The first Japanese chips were introduced into the US market in 1978. By 1980, Hewlett-Packard reported that Japanese chips had a failure rate *one-sixth* that of the highest-quality US competitor. Consequently, by 1986 Japan had managed to surpass the United States in the memory chip market. Japanese market dominance was secured largely through superior product performance, ample investment capital, and timely product delivery (furthered by a stable network of material suppliers) (Borrus et al. 1983).

By the mid 1980s, Japanese companies also controlled the production of semiconductor raw materials, including silicon and gallium arsenide. During the 1980s US firms still supplied about 90 percent of photolithography equipment, but that share is expected to drop below 5 percent by 2000 (Executive Office of the President 1994, p. 3). During the late 1980s, in an effort to regain lost market share, US firms initiated sweeping changes in both how and where they organized production. Large, vertically integrated companies sought to improve manufacturing methods and reduce the time it took to move chips to market, primarily through internal organizational changes. As Angel (1994, p. 109) found, firms have combined previously discrete functions of design, manufacturing, and assembly.

Known as "vertical disintegration," the process of breaking off various production steps commenced in the 1970s, when US companies, in an effort to slash costs, started shipping the least technically complex steps of chip manufacturing—assembly and testing—off to South Korea, Hong Kong, and the Philippines. Semiconductor packaging, in which the metal "leads" are attached to dies, soon followed (Scott 1988). Once established, the presence of semiconductor assembly in Southeast Asia created a skill base and work practices that promoted further migration of production functions from Silicon Valley. By 1988, US firms had refined chipmaking methods to a point where it became technically feasible for some to ship wafer fabrication to established assembly and packing regions in Asia.

Shrinking Geometry, Growing Cost

While external factors such as increased competition help account for restructuring, industry reconfiguration also is furthered by technological innovation. As was discussed in chapter 1, the industry's focus on achieving smaller chip geometries has pushed up the cost of equipment such as photolithography tools. In the past, the high cost of fab construction was counterbalanced by high-volume production and sales. Today, that logic has been overturned. Companies such as Intel are still willing to construct fabs to produce leading-edge microprocessors, in part because their products remain among the most complex to manufacture. In Intel's case, this business strategy yields rewards that are circular and cumulative: Intel's ability to overwhelm its competitors by outbuilding them gives it the sustained ability to bankroll the cost of fab construction in order to maintain its overwhelming market lead.

For other firms, particularly new ones specializing in the design of ASICs, costly wafer fabs are increasingly perceived as investments that deliver few returns. Some firms find that they can extract more value by focusing on customized chip designs, rather than on costly high-volume manufacturing.

Market Swings

As economic developers the world over try to lure wafer fabs, Silicon Valley has managed to eliminate most of them, in part as a way to avoid market volatility. While the high-volume chipmaking giants continue to operate advanced manufacturing facilities in Silicon Valley, it is unlikely that the region's former manufacturing base will return. Just as semiconductor manufacturing replaced heavy industry, chipmaking in Silicon Valley is being supplanted by a new, more stable growth engine. By 1997 the region was in the midst of its biggest boom in two decades, but the boom was based on software and Internet applications rather than silicon. Observers posit that the new structure of the Silicon Valley economy marks an effort to move away from the highly volatile semiconductor industry (Markoff 1997).

In contrast to industries driven by consumer demand, the dictates of Moore's Law make the chip industry more similar in nature to farming than manufacturing. Like farmers, chip firms each year grow more transistors and hope for the best. The same applies to fab construction. Andrew Grove told *Fortune:* "Our fabs are fields of dreams. We build them and hope people will come." (Kirkpatrick 1997, p. 63) As the number of chips on transistors has mushroomed from several thousand to several million transistors, the industry must hope for a continuous consumer appetite for new product applications.

Often the hunger for so much new transistor power has fallen short of the supply, resulting in plummeting prices. For example, between 1995 and 1996 the prices of DRAM chips dropped precipitously, causing chip revenues to fall by 40 percent and again by 14 percent in 1997. Such market swings can prompt firms to postpone or even cancel new fab construction.

For example, in 1997 Motorola announced that it was abandoning the unprofitable DRAM chip market altogether. That decision caused Motorola to change its plans for a recently completed plant near Richmond, Virginia (built in collaboration with the German firm Siemens AG). Initially, the White Oak semiconductor facility will be used by Motorola to make DRAM chips, according to an article in the 2 July 1997 *Washington Post.* However, Motorola will phase out production of the most common memory chip by 2000, after which it will use the facility to make other types of memory chips, including a higher-priced chip known as a fast static random-access memory. The White Oak facility, which received $19.9 million in incentives from the state, began limited DRAM production in 1998. The DRAM phase-out means that Motorola must buy new equipment and provide workers with additional training. When the plant reaches full production, around 2000, it could employ up to 1500 people.

Market volatility also has modified plans for another nearby Motorola fab in Goochland County. Originally, Motorola planned to build a new fab to make PowerPC microprocessors. The chips are largely used to power Apple's Macintosh line. However, a dramatic fall in the Mac maker's fortunes caused Motorola to scrap its original plans for the site. In good news for Virginia, however, the revised Goochland County facility will be double the size of the complex which was originally planned. Instead of making microprocessors at the site, according to the 4 December 1997 *Washington Post,* Motorola now plans to make a host of semiconductor products to power telephones, pagers, and other electronic devices. When complete in 1999, the $3 billion West Creek facility will be the largest of the state's three new semiconductor plants. The West Creek facility marks Motorola's largest single capital investment ever. The five-building, 1.5-million-square-foot complex will create 2500 jobs. Motorola officials told the *Post* that the plant ultimately could employ up to 5000. Though diversifying its manufacturing portfolio may reduce the manufacturer's vulnerability to market swings, volatility in any one segment of the industry may still cause Motorola to put its plans for the site on hold. A Motorola representative told the *Post:* "The computer chip industry is very volatile; our schedules change."

Owing to the industry's remarkable growth and the chip's increasing importance, it is easy to see how luring semiconductor manufacturers may appear to be a sound long-term development strategy. Indeed, as Intel's experience in Hillsboro, Oregon, shows, the 20-year presence of the company has provided a stable and growing source of jobs. Intel's longtime and expanding presence there has helped to lure additional chip makers, including Fujitsu and LSI Logic. However, some question New Mexico's enormous financial commitment to an Intel plant that will have a useful life of only 10–15 years (SWOP 1995).

It remains unclear whether wafer fabs alone are sufficient to spark a cluster of chip companies. The experience of the industry suggests that, at least in the initial stages, the most crucial component of sustained growth is research and development. In other words, Silicon Valley did not sprout from a wafer fab, but from innovation breakthroughs. The same is true of Motorola's Phoenix hub and the string of the Texas companies spurred by the longtime presence of Dallas chip maker Texas

Instruments. Intel's Oregon operations similarly house research, development, and marketing groups.

As part of Motorola's plans to revise its planned PowerPC fab near Richmond, Virginia, the facility also will become the site of Motorola's most advanced chip research and development functions. Historically, research and design for the Illinois-based manufacturer is housed near Motorola's Phoenix and Austin manufacturing hubs.

As the industry moves away from established regions, companies may emulate Motorola's decision to move research and design functions to one of its two new fabs near Richmond. Even if the research and design functions fail to follow, it is true that companies can extend the longevity of plants by retooling fabs that have become obsolete. But just how long such plants will remain viable is an open question. It is not clear what will happen to chip manufacturing plants, workers, and government coffers, when manufacturers ultimately find themselves unable to squeeze more transistors on to silicon.

To summarize, both market factors and the challenges of making smaller, faster, defect-free chips have contributed to the industry's recent reconfiguration. Since the chip's inception, US firms have moved from a position of unbridled hegemony to fierce competition split roughly between American and Far Eastern firms. As the technology matured, Japanese firms were able to replicate and surpass American methods by focusing on product quality and delivery times, instead of merely on continuous innovation. As Japan began to erode the United States' share in the market for semiconductors, it also established a firm hold in the production of materials such as silicon, gallium arsenide, and wafer fabrication equipment.

Such competitive challenges have led to large-scale organizational and geographic shifts among US firms. Texas Instruments once designed its own manufacturing equipment, made silicon ingots and silicon chips, and assembled and packaged them, but today's Silicon Valley upstart may seek to differentiate its bundle of transistors by offering unique customized designs. "Fabless" firms leave chip manufacturing, assembly, and packaging largely to offshore subcontractors.

Combined, such challenges have prompted the industry to move from a homogeneous business model that finances continuous innovation through high-volume production to a series of different market adaptations, including agreements to license intellectual property and subcontracting relationships. Firms such as Intel that continue to manufacture cutting-edge products have reorganized operations to minimize the uncertainties of production, far in advance of production, thus making it possible to build new, state-of-the-art fabs in far-flung places.

New Business Models

Semiconductor firms have entered into formal operating agreements with one another since the industry's inception. To ensure supplies during periods of peak demand, US firms have obtained "second source" agreements with other companies. Firms also have long contracted assembly to offshore suppliers. What distinguishes the current round of alliances is the degree to which the shifts involve the chemical-intensive phase of fabrication. According to one study, firms entered into more than 1000 strategic agreements during the 1980s (Kogut and Kim 1991). Moreover, the new cooperative manufacturing ventures, akin to the variable production processes used to make chips, are in constant flux. The terms of most formal agreements are under 5 years. Others may be even shorter, depending on market conditions and the ease with which chip foundries are able to fulfill designers' orders.

The following discussion examines three types of strategies firms have adopted to improve competitiveness: fablessness; strategic partnerships; and free-standing production. The six public companies selected here represent to varying degrees how firms have sought to improve competitiveness. The sample includes two fabless companies and four companies that continue to construct and operate wafer fabs. The six companies manufacture three major categories of semiconductors, including microprocessors, memory chips, and application-specific devices. The product portfolios of the six companies are largely consistent with the business strategies adopted. That is, fabless firms typically make application-specific products. Microprocessor manufacturers largely continue to operate their own fabs. The variation is reflected in size, as measured in terms of total revenues. Intel's sales in 1996 came to more than $20 billion. In the same year, the revenues of the fabless company SEEQ Technol-

Table 3.2
Three types of production. Source: US SEC 1997.

Company Headquarters	Partner	Fab locations			
		United States	Asia	Europe	Middle East
Fabless Companies					
SEEQ Technology	American Microsystems	Idaho	Korea		
	Hualon Microelectronics		Taiwan		
California	International Microelectronic Products		Japan		
	Ricoh				
	Rohm				
	Samsung				
Cyrix[a]	IBM Microelectronics	New York		France	
Texas	SGS Thomson	New Jersey		Italy	
		Vermont			
		Arizona			
		Texas			
Strategic Partners					
Advanced Micro Devices	Sony Microelectronics	California	Japan	Germany	
	Fujitsu	Texas			
California					
National Semiconductor	Intel	Maine		Scotland	
	National Semiconductor Sunrise	Texas			
California	Synaptics	Utah			
	Integrated Information Technology				
Freestanding Fabs					
Intel	National Semiconductor	California		Ireland	Israel
California	Advanced Micro Devices	Arizona			

Table 3.2
Continued

		Fab locations			
Company *Headquarters*	Partner	United States	Asia	Europe	Middle East
		New Mexico			
		Oregon			
		Puerto Rico			
Cypress Semiconductor *California*		California			
		Texas			
		Minnesota			

a. Purchased by National Semiconductor in 1997.

ogy were around $31 million. Like SEEQ, Intel's competitor Cyrix also focuses on design rather than manufacturing. In contrast to SEEQ, which designs chips for local-area networks, Cyrix designs microprocessors. National Semiconductor, which purchased Cyrix in 1997, has a more diversified production portfolio. One feature common to five of the six companies is that, while their wafer fabs are increasingly located in other parts of the United States and the world, most maintain corporate headquarters as well as some experimental production facilities in California's Silicon Valley.

The three manufacturing strategies (fablessness, partnerships, and freestanding fabs) are differentiated primarily in two ways. First, firms are distinguished by the degree to which manufacturing remains integral to the company. Table 3.2 represents by rows in descending order the degree to which firms have reorganized production. SEEQ and Cyrix do not make chips; they rely on third-party suppliers. Advanced Micro Devices and National continue to manufacture microchips but share product development methods. AMD and Sony share manufacturing lines. Finally, largely "free-standing" companies including Intel and Cypress primarily obtain supplies from their own fabs. Intel has at various times subcontracted 20–30 percent of its production out to third-party suppliers (Hayashi 1988, p. 36). Though both Intel and Cypress maintain corporate headquarters and research functions in California, their fabs are increasingly located outside Silicon Valley. Indeed, this characteristic is common to all three manufacturing strategies.

Three Types of Production

As the three categories of production suggest, each configuration also is distinguished by the degree to which companies share sensitive information. In general, free-standing firms share the least information, strategic partners the most. Fabless firms, which license third-party suppliers to manufacture chips, fall somewhere in between.

Fabless Firms

The first type of reconfiguration is the most dramatic. As the name suggests, fabless companies design chips but rely completely on third-party suppliers for manufacturing. In contrast to high-volume producers, such as IBM, Motorola, and Intel, fabless companies design customized products that suppliers then manufacture in small, variable batches. In a world increasingly saturated with transistors, fabless companies have carved out a market niche in which they simply design ASICs. In contrast to microprocessors, which require more than 300 individual manufacturing steps, chips designed by fabless companies are easier to manufacture. One result is that fabless companies help utilize the growing glut of existing manufacturing facilities too old to produce state-of-the-art chips. One California executive who closed his fab reasoned: "It costs $9 million to run a fab. I can buy that value of fab for less than $4 million." (SEEQ 1991) The glut of older fabs is an even greater incentive to go fabless than the excess of transistors.

The primary objective of most fabless companies is to design the best custom product in the shortest possible time. Fabless firms then transmit product specifications to subcontractors charged with fabricating the chips. Some suppliers design and make their own chips as well as fill orders for fabless firms; others (those called "foundries") manufacture but do not design chips. Most foundries are located offshore, a continuation of the assembly functions farmed out by US firms years before. In a Dataquest study of 137 new semiconductor startups from 1977 to 1985, 43

US companies were found to have more than 70 agreements with Asian companies (Rice 1987). Almost all fabless companies, as well as firms that still make their own chips, have foundry partnerships or second-source agreements with Asian companies.

Although most chips designed by fabless companies require fewer manufacturing steps than microprocessors, Texas-based Cyrix followed the fabless model to compete with Intel, which currently ranks seventh among companies in global production capacity. Like the smaller fabless firms, Cyrix sought to focus on product design, market development, and customer support rather than on developing process technologies and operating manufacturing facilities.

Since 1990, French-German SGS-Thomson Microelectronics has manufactured chips designed by Cyrix and then sold them to Cyrix. Cyrix in 1994 also announced a five-year agreement to manufacture advanced microprocessors with IBM Microelectronics, based in Hopewell Junction, New York. Pursuant to the agreement, Cyrix made a capital equipment investment of about $88 million in an IBM manufacturing facility. Though the fabless strategy allowed Cyrix to devote more resources to research and development, it also made it harder for Cyrix to directly employ manufacturing processes comparable to Intel's (Cyrix 1996).

In addition to firms that decide not to operate fabs from the outset, some small companies have followed the lead of fabless startups. SEEQ Technology closed its fab at Fremont, California, in 1992. In 1996, SEEQ designed chip sets for local-area networks and had manufacturing arrangements with six semiconductor manufacturers: American Microelectronics, Hualon Microelectronics, Ricoh, Rohm, Samsung Semiconductor, and Taiwan Semiconductor Manufacturing (SEEQ 1997a, p. 11).

The proliferation of fabless startups and the closing of fabs have led to the establishment of a trade organization, the Fabless Semiconductor Association, whose members include fabless firms and companies that continue to manufacture their own chips. In 1997 more than 60 members of the FSA obtained at least 75 percent of their chips from foundries (FSA 1997).

Beyond the FSA, the magnitude of fablessness is difficult to quantify, because four-digit SIC codes fail to distinguish between fabbed and fabless companies. The US Department of Commerce does produce a six-digit SIC code at the national level for "wafer suppliers," but such designations still fail to distinguish between firms that merely design chips and foundries that produce them.

Strategic Partnerships

The second outgrowth of restructuring in the industry involves large producers that have entered into agreements to cooperate rather than compete. In contrast to fabless partnerships between a designer and a producer, fab or product sharing partnerships typically involve companies that continue to develop and manufacture chips. The partnerships can require firms to share technology and often manufacturing facilities and costs or can simply consist of strategies not to compete in certain markets. For example, rather than compete with a giant like Intel, National Semiconductor repositioned itself to supply circuit boards and chips to that company (National Semiconductor 1995).

In some cases, firms negotiate complex legal agreements in order to both improve manufacturing methods and expand markets. For example, instead of attempting to win back market share from its Japanese rivals, Advanced Micro Devices sought to emulate Japanese manufacturing methods by agreeing to pool production space and technology with Sony Microelectronics. In 1990, as part of the deal, AMD sold its two San Antonio fabs to Sony. Sony, in turn, upgraded the plants and now uses one of them as a base to teach AMD executives and engineers about Japanese production technology (SEMI 1994; Florida and Kenney 1990). In exchange, Sony received from AMD access to certain proprietary technologies. In 1994, in order to gain greater access to markets overseas, AMD also entered into an agreement with Fujitsu Limited to build a $700 million flash memory chip fab in Aizu-Wakamatsu, Japan. The two companies completed a second facility at the Japanese site, which commenced operation in late 1997 (AMD 1998).

In cases where there is a clear market leader, it also may be in a firm's interest to share proprietary technology in order to manufacture uniform products that generate market expansion rather than the potential fragmentation that could occur from product differentiation. For example, in 1995 Intel and its chief competitor in the microprocessor market, AMD, negotiated a patent cross license agreement that includes the new instruction set that enhances the multi-media capacity of Intel's microprocessors (Slater 1996b). Developing a single instruction set will benefit software developers and ultimately lead to faster market growth. The agreement represents a departure for Intel, which historically has sought to compete on the basis of technological superiority. According to the *Microprocessor Report* analyst Michael Slater, the agreement signals Intel's confidence in its overwhelming ability to maintain market share both by being first to market and by having enough manufacturing capacity to simply overwhelm its two chief competitors in the microprocessor market.

Free-Standing Fabs
The third variant involves companies that continue to perform most of their own manufacturing in house but have reorganized internally to improve manufacturing methods and respond more rapidly to market signals. In some cases, internal reorganization enables companies to locate new high-volume wafer fabs near global markets. As Angel (1994, p. 109) notes, during the 1980s Intel, AMD, and IBM sought to reduce technological uncertainties in manufacturing and sales by integrating production and marketing into planning and design functions. The shift is particularly pronounced among microprocessor producers because technological uncertainties inherent in the development of these products give the firm an incentive to maintain control over production processes, rather than contract fabrication to foundries.

Andrew Grove explained in the *Harvard Business Review* why manufacturing matters to those who design and produce the most sophisticated chips: "In my almost 30 years in the semiconductor industry, I have learned that there is never enough production capacity to produce leading-edge products, and there is always more than enough capacity for yesterday's technology. . . . Just as you cannot build formula one racing cars with stock parts, you cannot develop high-performance microprocessors with generic semiconductor technologies." (Grove 1991)

W. J. Saunders III, chair and CEO of Advanced Micro Devices, once famously declared that "real men have fabs" (Weber 1991). Though it is clearly in the interest of some firms to fabricate chips, organizational changes aimed to reduce design time and product failure make it possible for companies to build new wafer fabs outside Silicon Valley. Angel

(1994) describes how, during the 1980s, high-volume microchip manufacturers moved from a linear organizational model to an integrated production model. In the past, product planning, design, fabrication, assembly, and sales were considered discrete steps. Reorganization allowed firms to integrate production concerns into planning and design. One result of the change is that companies now are free to build new fabs far from corporate headquarters. Reorganization also allows companies to condense the time between microchip development and high-volume production. As recently as 1993, it took 6–8 years to go from research to production. By 1997, the same process took 2–3 years.

AMD and Intel have adopted such organizational models. Sunnyvale-based AMD moved most of its high-volume operations from California to Texas during the late 1980s. In addition to the shift toward Texas, AMD in 1996 broke ground on its first European manufacturing facility, Fab 30, in Dresden. The 875,000-square-foot plant houses both a design center and a 90,000-square-foot clean room for the manufacturing of microprocessors and other high-volume products.

Intel, which now builds most of its high-volume fabs outside California, also maintains an experimental manufacturing facility near its Santa Clara corporate headquarters and a second such facility in Oregon. Intel operations are diversified through more than 30 countries around the globe. An Intel chip developed in Oregon may be manufactured in Ireland or in Israel, packaged and tested in Malaysia, then sold to a customer in Australia.

Economic geographers observe that in, most cases, the dispersal of production from Silicon Valley is not random but is motivated by the presence in other regions of an existing technical base of electronics producers, workers, and suppliers. As was noted in chapter 1, new fabrication facilities in the United States have tended to cluster in the Pacific Northwest and the Southwest. Within the United States, firms are primarily drawn to states with an established semiconductor manufacturing base, such as Texas, Arizona, New Mexico, and Oregon. (See tables 1.2 and 1.3.) For example, Motorola has long operated a manufacturing hub near Phoenix. Dallas-based Texas Instruments helped to spur the Lone Star State's recent chip plant boom. Intel, which is expanding operations in Phoenix and Texas, built its first chip plant in Oregon more than 20 years ago and now has five fabs clustered in the Hillsboro area. Other companies, including Fujitsu, LSI Logic, and Hyundai, have recently announced plans to build fabs in Oregon.

Relocation is mirrored across the Pacific, where chipmaking is shifting from Japan and South Korea to Taiwan and Singapore. Between 15 and 20 new megafabs are slated for construction in HsinChu, a small city in northern Taiwan. In 1995, Taiwan produced 3 percent of the world's chips. As engineers there become increasingly familiar with fabless firms' production methods, Taiwan will likely become a dominant force in chipmaking by 2000 (LaPedus 1996).

Like Taiwan, Singapore—originally a low-cost test and assembly location for US and Japanese firms—has experienced robust industry growth. The country is home to 35 multinational and indigenous semiconductor companies. In 1996 the Singapore Economic Development Board reported that the semiconductor industry represented nearly 14 percent of Singapore's total electronic industry. AMD, NEC Semiconductors, Matsushita Denshi, Fujitsu, SGS-Thomson, Siemens, and Sony now design integrated circuits there (EDB 1996a). Singapore also supports sizable foundries, such as Chartered Semiconductor Manufacturing Inc. In 1997, *Semiconductor International* named that company's Fab 2 one of the top fabs in the world, making it the first foundry to earn such a distinction (CSM 1997). Chartered's Fab 2, completed in 1996, is a fully automated Class 1 facility with robotic controls that allow wafers to remain in an ultra-pure manufacturing environment throughout the production cycle. According to the Economic Development Board of Singapore, Chartered's achievement is just the beginning. By 2005, there will be approximately 20 more wafer fabrication plants in that country (EDB 1996b).

Though organizational moves to minimize uncertainties associated with chip production certainly enable firms to construct new fabs outside Silicon Valley, other factors have helped fuel the exodus. As mentioned, the decline of chip manufacturing there was driven by market volatility. However, Silicon Valley, to some degree, also is a victim of its own success.

In a 1993 survey by Integrated Circuit Engineering, an industry research group, semiconductor firms indicated that rising production costs made Silicon Valley a less desirable place to live. Chip firms cited limited land supplies for new factories and worker housing, as well as complicated regulatory processes and worker compensation problems as factors in their decisions to build wafer fabs outside Silicon Valley (ICE 1993). The lesson was hammered home when Intel announced its decision to construct a costly new fabrication facility at an existing operation in New Mexico.

Just as clustering provides benefits by pooling expertise, equipment, and infrastructure, industrial concentration can create economic "negative externalities." Among the most commonly cited negative externalities in Silicon Valley are pollution, traffic congestion, and crime (Saxenian 1981).

By most accounts, Silicon Valley's leading US contender is Austin (*Economist* 1997, p. 14). No newcomer to microchip manufacturing, the Lone Star State has long been home to firms such as Texas Instruments. By 1995, the Austin region employed 26,000 people in the semiconductor industry. In addition to semiconductor companies, Austin also supports about 250 computer and computer-related companies, which together employ another 38,000 people.

Texas attracts about 6000 Californians each year with affordable housing, cleaner skies, and fewer hours spent on clogged freeways. In Silicon Valley the average computer professional's house may cost $500,000; in Austin a comparable residence sells for about $150,000 (*Economist* 1997, p. 14). According to a 1997 report in the *San Jose Mercury News*, Santa Clara County is the most expensive housing market in the United States. In addition to pulling professionals from Silicon Valley, Austin recently lured Samsung Electronics to locate its first US wafer fab there. Among the other chip makers with fabs in Austin are Motorola, AMD, Cypress Semiconductor, Crystal Semiconductor, and Sematech. Intel also recently announced plans to build a new fab in Texas.

Amkor Technology, a large Pennsylvania-based foundry with offices in Santa Clara, illustrates how the trend toward fabless production increasingly prompts US firms also to look overseas for chip supplies. In 1997, the *Silicon Valley Business Journal* reported that Amkor announced plans to construct three wafer fabs in Korea. In its new role as a wafer manufacturer, Amkor will be competing with Taiwan Semiconductor Manufacturing and with Singapore's Chartered Semiconductor Manufacturing.

Amkor's decision to build abroad was motivated by both environmental regulations and labor cost. One industry expert told the *Business Journal*: "You can't build a fab in California because of all the environmental restrictions." (Barlas 1997, p. 5) Amkor's company spokesman told the *Business Journal* that the company selected Korea as a production site because of the country's ample water and labor supplies. Amkor plans to devote at least 40 percent of its new plant's production to wafers for Dallas-based Texas Instruments, with which Amkor signed a technology licensing agreement a year before the Korean fab was completed.

Summary

Economic restructuring has resulted in movement away from fairly uniform business models toward a heterogeneous system of production styles. How a firm chooses to produce chips is dictated in part by the firm's manufacturing requirements. Firms that plan and design highly differentiated products, such as application-specific chips, are often fabless. In other cases, companies have an incentive to promote product uniformity and thus expand markets by entering into strategic partnerships. Former competitors also have increasingly teamed up to gain manufacturing skills and expand markets. Producers of hard-to-make microprocessors still have an incentive to invest in costly new wafer fabs. However, internal efforts to integrate production concerns into design have enabled large US manufacturers to locate sophisticated fabs in foreign countries.

The primary result of such shifts is that formerly clear-cut distinctions such as "firm" and "industry" are no longer adequate to characterize what is increasingly becoming an interdependent network of product designers, producers, and assemblers. Not all semiconductor companies designated as manufacturers by standard industrial codes continue to operate fabs, and fabs formerly owned and operated by US firms may be built or operated in tandem with foreign companies.

The reconfiguration away from free-standing firms is accompanied by geographic shifts of new high-volume fabs to the southwestern and northwestern areas of the United States and to East Asia. In most cases, firms build new fabs in regions with an existing high-tech manufacturing base.

There are signs that the industry increasingly is looking to build fabs in the southeastern United States, as evidenced by the completion of two fabs in Virginia.

At first glance, establishing a manufacturing base organized around chip production appears to be a sound economic development strategy. As the experience of Silicon Valley shows, however, chip production has environmental and economic downsides. Among these are pollution, volatile market swings, and longer-term uncertainties. Just as firms sought to build new fabs away from California, there is no guarantee that the sites under construction today in other parts of the United States and the world will still be in operation by the year 2020.

4

Environmental Challenges of Restructuring

Restructuring has helped some US producers to regain market share by enhancing both design and manufacturing methods. One downside of the industry's economic resurgence is that environmental effects that can accompany restructuring and globalization are difficult to pinpoint and evaluate. The newfound ability of companies to build outside established chipmaking regions such as Silicon Valley gives companies enormous leverage over cities, states, and developing countries anxious to lure manufacturing jobs and tax revenues. The footloose character of fab construction may make pollution harder to understand and to manage. The Toxics Release Inventory best illustrates how geographic factors complicate efforts to assess the industry.

Using data developed for firms in chapter 3, the following case study first shows how the TRI fails to account for the effects of globalization and economic restructuring in the semiconductor industry. The TRI is based on the increasingly outmoded assumption that production still is conducted by one firm under one factory roof in the United States. Although TRI data show a decrease in toxics released and transferred by the semiconductor industry since 1988, it is likely that organizational and geographic shifts have contributed to the declines.

Another problem with relocation is that, although federal laws help to ensure that states follow uniform environmental standards, not all jurisdictions possess the same level of resources, industry-specific expertise, and even political will to administer and enforce federal laws. The uneven capacity among states and countries to administer environmental programs may help to heighten the controversy associated with industry expansion.

Misleading Toxic Release and Transfer Trends

In the firm's annual report for 1995, National Semiconductor's management pointed to an 88 percent reduction in chemicals reported to the Toxic Release Inventory since 1988 as an example of environmental progress (National Semiconductor 1995). The Semiconductor Industry Association similarly cites TRI data since 1988 as an indication that toxics released and transferred by the industry are declining (SIA 1997). However, what National Semiconductor and the SIA fail to discuss is whether the declines are due to superior environmental performance or to other factors. As the following case study makes clear, at least some of the declines are due to globalization and economic restructuring.

The chief reason that the TRI reports fail to account for the effects of globalization and restructuring is that organizational change in the industry has outpaced federal record-keeping methods. In a supplement on Silicon Valley, *The Economist* (1997, p. 19) observed: "It is hard to exaggerate how far ahead of the American government Silicon Valley has moved. Even the statistics are a quagmire."

Established in the 1930s, the Standard Industrial Codes fail to reflect some of the most basic features of high-tech sectors. For example, software publishers in California are ineligible for research and development tax credits because the SIC fails to classify them as "manufacturers" (Hamilton 1996). The SIC is currently undergoing an overhaul to make it more accurately reflect the US business landscape. Yet it remains unclear whether the revised codes will precisely reflect the new industrial configuration of chipmaking. Absent correct statistics, information to help chart economic alignments among designers, foundries, and manufacturers is best obtained through trade press articles and reports filed annually by public companies with the Securities and Exchange Commission.

The weaknesses of the SIC system are incorporated into federal environmental tracking tools such as the TRI, which identifies sectors, industries, and individual facilities with two-digit and four-digit SIC codes. Congress created the TRI as part of Title III of the Superfund Reauthorization Amendments in response to the deadly 1994 gas leak in Bhopal, India. Title III develops emergency response provisions in order to avert similar tragedies. The EPA developed the TRI to provide the public with continuing information on firms' emissions and transport activities.

The TRI is a useful tool for people who use the database to track chemical emissions at facilities that produce items designed and sold by one company. Interested individuals may use the database to track the 343 chemicals (expanded to 643 in 1994) on which firms must report. For example, a Silicon Valley public interest coalition staged a protest at a Santa Clara IBM plant in 1989 after learning from the TRI that the facility was responsible for emitting the largest quantity of chlorofluorocarbons in the United States. The group demanded and received from IBM a pledge to reduce and eventually phase out the use of the ozone-depleting solvents (US EPA 1990, p. 323).

Although the tool helps individuals to track and monitor progress at facilities owned and operated by a single firm, it is a less useful measure to gauge progress in reducing TRI transfers and emissions in the chip industry. For example, the TRI cannot be used to hold a fabless company accountable for the products it designs and sells, nor is it a useful tool with which to compare performance among chip firms. Fab closure, strategic alliances, and the increasing ability of chip makers in recent years to build new fabs abroad further limit the TRI's usefulness as a measure of industry performance. Trade press articles and company shareholder reports help to plug some data holes, but notable gaps remain. The gaps are more glaring for product identity and output data—information necessary to evaluate environmental performance among different facilities and firms.

Loose Lips Sink Chips

Chapter 1 shows how a chip's value is derived not so much from the silicon, water, and chemicals used to make it as from information about how the chip is designed and manufactured. Though companies jealously attempt to guard sensitive information through patents and confidentiality claims, the case of Shockley's Traitorous Eight shows how difficult chip-making knowledge is to contain. The eight engineers who defected from Shockley helped to lay the foundation for many of today's major chip companies. Although most industries invoke patents to protect information, evidence of the rapid dissemination of semiconductor technologies suggests that patent system protections in this industry are weak (Lamond and Wilson 1984, p. 46).

The "leaky" nature of industry information is both a boon and a bane. For example, when specialists defect to start their own firms, large chip companies lose valuable assets. However, the spinoffs formed by defectors from large firms have traditionally represented an important source of industry innovation. Spinoffs—whether in Silicon Valley or in Scotland's Silicon Glen—are an importance source of innovation, but also an investment loss to large companies, which lose skilled workers and information. Spinoffs ostensibly represent an additional source of competition to established chip companies.

Despite the highly secretive nature of the industry, a great deal of information regarding companies' product plans is compiled by third parties and published for free or for sale in the trade press and in public government sources, such as the Edgar database maintained by the US Securities and Exchange Commission. Trade publications also provide a wealth of information, some of which companies claim is confidential. For example, Intel typically conceals the specific identity of chips made in its individual facilities. Yet unconfirmed data on product identity is readily available in trade reports. For example, according to *Microprocessor Report*, Intel's Fab 12 in Arizona most likely manufactures a mid-range Pentium or P55C chip with MMX multimedia extensions that enhance video display (Slater 1996a). Intel's two key competitors in this market are Cyrix's M2 and Advanced Micro Devices' K6.

Furthermore, the increasing interdependence of chip firms makes it difficult to establish the legitimacy of at least some confidentiality claims. For example, it is well established that Intel and one of its chief competitors, AMD, have entered into an agreement to share MMX platform technology used at the XL Project site and other Intel facilities (Slater 1996c). The title of a recent tome penned by Andrew Grove may help to explain why Intel prefers to remain so secretive. *Only the Paranoid Survive* (1996) charts how Intel lost important markets and later relied on "paranoid" management strategies to achieve its current unrivaled status.

Paranoia regarding production data is common even among fabless firms. The lack of public information on third-party suppliers makes it virtually impossible to know whether a fabless firm designs products with environmental criteria in mind. For example, although Cyrix identifies its chief foundry partners as SGS Thomson and IBM, the annual reports filed by Cyrix with the Securities and Exchange Commission (e.g., Cyrix

1996) do not specify where the fabs of its third-party suppliers are located or how much the fabs produce. The fabless producer SEEQ similarly identifies its foundry partners in its annual reports, but the reports do not specify where the foundry partners' fabs are located or how many chips they supply to SEEQ. According to SEEQ's 1996 Annual Report (1997a, p. 29), "a substantial number of the Company's products are manufactured, and all of the Company's products are assembled, by independent foundries and assembly suppliers in foreign countries, including Taiwan, Japan and South Korea." In addition to trade secrecy concerns, it is likely that fabless companies such as SEEQ fail to specify exact locations of suppliers and quantities in public reports not to deliberately conceal sensitive information but simply to save on printing costs: production relationships with foundry partners can change very rapidly.

Most foundry agreements are for 5 years and can be modified or terminated quickly. For example, SEEQ's 1996 Annual Report (1997a, p. 28) states: "The Company generally does not have long-term, non-cancelable contracts with its wafer suppliers. Therefore, the Company's wafer suppliers could choose to prioritize capacity for other uses or reduce or eliminate deliveries to the Company on short notice."

To gain further insight into the possible sources of the TRI declines, the following discussion pairs economic information developed for six US companies in chapter 3 with the data reported by the companies to the TRI. Recall that the six companies range in size from small ones such as SEEQ to giants such as Intel. In fiscal year 1996, SEEQ's revenues were $31,338,000 (SEEQ 1997a, p. 2). Intel's sales in that year came to $20.8 billion (Kirkpatrick 1997, p. 60). The distinction is important because it is generally the case that smaller companies report that they are less able than Fortune 500 firms to devote resources to environmental, health, and safety issues. Combined, the six companies account for between 5 percent and 12 percent of all releases and transfers reported to the TRI by semiconductor firms operating in the United States between 1988 and 1995.

Table 4.1 shows emissions and transfers off site of toxic chemicals for five firms on which TRI data are available for 1988 and/or 1995.[1] Overall, between 1988 and 1995, releases and transfers fell for three of the four

---

1. Though SEEQ filed reports up to 1990, no TRI data are available for 1995, because SEEQ had permanently closed its California fab by then.

Table 4.1
TRI releases and transfers (pounds). Source: US EPA 1997b. (Computer search conducted for facilities that use 3674 as primary SIC code. Search used medium level of emissions detail. Searches for "high" or "low" levels of detail produce slightly different totals.)

	Reporting year	
	1988	1995
SEEQ Technology	2,950	NA[a]
Cypress Semiconductor	19,123	579
Advanced Micro Devices	442,246	308,670
National Semiconductor	580,861	209,917
Intel	523,692	2,618,082
Total	1,568,872	3,137,248

a. Not available.

firms for which data are available. Total TRI amounts increased between 1988 and 1995 only for Intel. That corporation's increases are likely due to significant expansion at reporting facilities such as those in New Mexico that occurred during the period, as well as to the EPA's decision to nearly double the number of chemicals on which firms are required to report.

To help account for TRI declines, table 4.2 illustrates changes during the same period in the number and location of facilities operated by five of the six companies for which TRI data are available. Recall that Cyrix is a fabless producer that contracts manufacturing out to IBM Microelectronics and SGS Thomson. The most obvious trend among the five firms is the shift of manufacturing facilities away from California to other parts of the United States. The number of AMD California facilities reporting to the TRI dropped from five to one over the period. Intel's dropped from three to one. SEEQ closed its California fab in 1992, and California-based Cypress opened fabs in Texas and Minnesota. As was discussed in chapter 3, the remaining AMD and Intel facilities in California are used primarily for research rather than for high-volume production and thus are not likely to trigger TRI reporting thresholds.

With the exception of Cypress Semiconductor, which experienced no net change in the number of reporting facilities during the period, the

Table 4.2
TRI facilities reporting. Sources: US EPA. 1997b; US SEC 1997. (A TRI "facility"
is not necessarily a wafer fab.)

	1988	1995
Cyrix	NA[a]	NA
AMD	California (5)	California (1)
	Texas (2)	Texas (1)
Cypress	California (1)	Texas (1)
	Texas (1)	Minnesota (1)
Intel	Arizona (2)	Arizona (1)
	California (3)	New Mexico (1)
	New Mexico (1)	Oregon (1)
	Oregon (2)	Puerto Rico (1)
National Semiconductor	Arizona (1)	Maine (1)
	Connecticut (1)	Texas (1)
	Maine (1)	Utah (1)
	Texas (1)	
	Utah (1)	
SEEQ Technology	California (1)	NA
Total	23	11

a. Not available.

number of facilities reporting to the TRI also declined. For all five firms,
the number of facilities that filed TRI reports fell from 23 to 11 from
1988 to 1995. AMD's fell from seven to two and Intel's from seven to
four. National's dropped from five to three.

There are a number of possible explanations for the TRI declines. One
is that the facility continued to operate but did not use any of the 643
chemicals in sufficient quantity to trigger a report. Another is that the
facility was temporarily idled or converted to another use. Finally, facilities may simply have been closed. In the case of National Semiconductor,
some of the TRI declines are clearly due to the closure of manufacturing
facilities. According to annual reports and features in the trade press, National Semiconductor in 1988 announced plans to phase out its Santa
Clara wafer production lines and shift production to Arlington, Texas.

National also sold, then leased back its Texas fab and its Santa Clara chip development center in 1990. The firm's Danbury and Tucson facilities were idled by 1994. During the period, National also sold its Puyallup fab to Matshushita, a Japanese microchip manufacturer. While National closed, sold, or idled some facilities in the United States, it also expanded and constructed new wafer manufacturing facilities overseas. National operates three fabs in Scotland's "Silicon Glen." According to company reports, National has maintained a manufacturing presence there since the 1970s (National Semiconductor 1997). However, National Semiconductor also significantly expanded production there during the reporting period. National reported that production at its facility in Greenock, Scotland, increased 85 percent in 1994 (National Semiconductor 1996). Though obviously not required to report emissions from its overseas facilities to the TRI, National Semiconductor reports that emissions at the Greenock fabs in 1994 declined more than 18 percent, and water use by 25 percent (ibid.). National also operates manufacturing facilities in Malaysia. More recently, National launched its first joint venture, National Semiconductor Shanghai Sunrise Ltd., in China (National Semiconductor 1997). That joint venture includes manufacturing and testing facilities.

Advanced Micro Devices similarly shut down all its manufacturing facilities in Silicon Valley, retaining only its experimental manufacturing facility. AMD also sold two of its Texas fabs to Sony Microelectronics. AMD currently operates a fab in Japan in conjunction with Fujitsu. It completed another fab in Japan, and one in the former East Germany.

Though National and AMD still operate fabs, in 1991 SEEQ announced plans to phase out all production at its California facility and to close the plant in 1992. SEEQ now obtains chips from a changing set of foundry suppliers. In 1996, SEEQ's manufacturing partners included AMI Semiconductor, Ricoh, Rohm, Samsung, and TSMC. According to its company reports, Ricoh lists manufacturing facilities in California and Georgia. Samsung operates wafer fabs in Korea and Portugal and recently announced plans to construct a new fab in Austin, Texas. The Taiwan Semiconductor Manufacturing Corporation operates two fabs in the HsinChu region, Taiwan's version of Silicon Valley, and has announced plans for two additional facilities in Taiwan.

SEEQ's foundry partners in 1995, the most recent year for which the EPA has released TRI reports, included AMI Semiconductor, Hualon Microelectronics, International Microelectronic Products, Ricoh, Rohm, and Samsung (SEEQ 1996). According to an electronic mail reply from a SEEQ representative, the company is too small to mount an active program to monitor the environmental performance of wafer suppliers. However, "SEEQ works under the assumption that local laws and ordinances in suppliers' respective jurisdictions govern day to day factory discharge and disposal" (SEEQ 1997b). SEEQ representatives said that the company does keep on file certificates of conformance as required for the non-use of chlorofluorocarbons or other ozone-depleting materials.

Though it is not possible to use the TRI to gauge emissions for SEEQ's overseas suppliers, releases and transfers in 1995 do appear on the database for SEEQ's US supplier, AMI. That year, releases and transfers for AMI's Pocatello, Idaho facility came to 18,202 pounds. However, it is impossible to determine what proportion of AMI's releases and transfers were due to production for SEEQ. It may be that SEEQ's other foundry partners have facilities in the United States as well but either failed to use TRI chemicals or used TRI chemicals in insufficient quantities to trigger reports. Another possibility is that TRI reports are listed under another SIC code. For example, SEEQ supplier Ricoh also makes photocopiers and other electronic equipment that appear under other parts of the SIC 36 designation and possibly also parts of SIC 35. In any case, it is impossible to use the TRI to identify what proportion of releases and transfers from SEEQ's third-party suppliers were due to production of SEEQ products. If revenues are a proxy for output, however, it is likely that environmental impacts from SEEQ are relatively small relative to those from Intel and National Semiconductor.

Similarly, no TRI reports exist for Cyrix. Like SEEQ, that Texas-based microprocessor firm is fabless. Before being purchased by National Semiconductor in 1997, Cyrix primarily obtained supplies from IBM Microelectronics and SGS Thomson. IBM Microelectronics reported emissions and transfers to the TRI for its US facilities in 1995, but it is impossible to determine how much of IBM's emissions are due to production for Cyrix. According to an electronic mail reply from IBM's Corporate Environmental Affairs department, it is impossible to segregate TRI release and transfer data for Cyrix's product because IBM manufacturers the product in the same buildings, using the same equipment as it does for the manufacturing of IBM's own product. According to the company's Corporate Environmental Affairs department: "If IBM performed foundry work in a building on production lines dedicated solely to Cyrix production, segregating TRI data would be relatively simpler." Recently, however, IBM spokesmen reported that implementing a scheme to segregate releases and transfers from its own products and those of Cyrix would be costly because the chips that the company manufactures for Cyrix's products represent only a small portion of the facility's production. Because production for Cyrix forms a relatively small proportion of output and TRI releases and transfers, IBM believes that environmental health and safety resources are better invested in "producing real environmental benefits" (IBM 1997).

In terms of industry-wide trends, it is possible that a proportion of the TRI declines since 1988 result from firms shifting production to newer, more efficient fabrication facilities outside California. In other words, emissions fell because one new fab can do the same work as two or three older fabs. However, in order to compare emissions among fabs, adjustments must be made for output, wafer type, and chemical use—data that, for the most part, are impossible to obtain.

Tracking Global Emissions

Though some firms merely shifted production away from California to other parts of the United States, since 1988 three of the six firms examined also constructed new fabs or expanded existing fabs abroad. Because firms with operations outside the United States and its territories are not required to report to the TRI, such geographic shifts of both production and pollution may contribute to the appearance of improved environmental performance. For example, by 1995 Intel had expanded its fab operations in Ireland and Israel, National had expanded production at its Scotland fabs, and AMD had constructed a new facility in Japan in conjunction with Fujitsu.

In 1993, to help the public track emissions of its non-US facilities engaged in computer and microchip manufacturing, IBM modified its internal reporting requirements. Though the EPA does not require US firms to report on emissions from overseas facilities, IBM voluntarily applied the TRI format to report on emissions from facilities abroad. The data—which do not distinguish between computer and microchip facilities—are available in the firm's 1996 annual environmental report for emissions from IBM facilities in the United States and abroad during 1995 (IBM 1996). That year, IBM's US sites, which manufacture chips and computers, used 27 chemicals in sufficient quantity to trigger reporting. Combined, the facilities released or transferred off site as waste 12.43 million pounds. That same year, total Superfund Reauthorization Amendments Section 313 and Pollution Prevention Act reportable quantities from IBM's non-US sites, including chip fabs, amounted to 23.4 million pounds (ibid.).

Between 1988 and 1995, several of the six firms studied here also sold or leased manufacturing facilities to foreign companies. For example, AMD sold two Texas fabs to Sony, and National sold a Washington fab to Matshushita. National sold its Arlington facility to an unspecified company and then leased the site back.

It is possible to use the TRI to track changes in facility ownership that result from lease or sale because the EPA supplies each reporting firm with a facility identification code that does not change when a plant is sold. For example, it is possible to use the facility identification code to determine that the facility in Puyallup once operated by National is now operated by Matsushita. The TRI facility identification code also makes it possible to monitor emissions from the Texas fabs sold by AMD to Sony. Though it is possible to use the TRI to continue to track emissions, the ownership transfers nonetheless contribute to the appearance that National's and AMD's company-wide (as opposed to facility) TRI emissions and transfers declined.

In view of the increased complexity of production relationships among chip firms, the TRI is an inaccurate measure of industry and company trends. As the six firms show, the development of the TRI coincides with the idling, closure, and geographic shift of fabs.

In addition to serving as a tracking device, the TRI is used by interested citizens and regulatory agencies trying to improve the chip industry's environmental performance. When a fabless company has foundry partners in the United States, it is possible to use TRI reports to examine emissions and transfers from foundries. However, few foundries make chips exclusively for one partner. For example, IBM produces chips for Cyrix as well as for IBM. Since fabless companies seldom pinpoint where their foundries produce chips and in what amount, it is impossible to use the TRI to track most emissions due to US foundries. Foundry agreements and the location of new US-owned fabs overseas further complicate efforts to assess environmental performance because US firms are not required to report on operations of overseas facilities. IBM's move to adopt the TRI format to report on its non-US facilities is a notable and promising exception. National Semiconductor and Intel supply information on the environmental performance of their overseas facilities, but not in a format identical to the TRI. It also must be noted that, although these three companies have elected to voluntarily report on some aspects of environmental performance of their foreign facilities, the data cannot be verified independently. What is clear is that for four of the firms examined here declining TRI numbers may be due in part to the expansion of US-owned fabs overseas as well as to fabless partnerships with offshore foundries.

Accountability

As production relationships among chip companies become more complex, so do the associated environmental issues. One problem is that fablessness may reduce the incentives for US companies to improve the environmental performance of their constantly shifting sets of foundry suppliers.

The rising prominence of suppliers is not confined to the semiconductor industry, but occurs everywhere from the automotive to the entertainment industry. To improve the environmental performance of suppliers, some firms, academicians, and environmental managers have endorsed the adoption of voluntary worldwide environmental standards. Among the most prominent is the 14000 initiative by the International Organization for Standardization (widely known as "ISO"), a private nonprofit organization that seeks to develop more uniform international business methods. The certification is predicated on the idea that attention to environmental concerns may become a source of advantage for suppliers and sellers.

The ISO 14000 series consists primarily of voluntary housekeeping measures that are not subject to certification. To date, companies must obtain certification through third-party audits for only one part of the series: ISO 14001, which describes in general terms features that a firm's environmental management system must contain (ISO 1995). For example, companies should be committed to "continual improvement and prevention of pollution" (ibid., p. 8) and should develop procedures and plans to identify the environmental aspects of different corporate operations.

What is potentially confusing about the ISO terminology is that the certification standards refer not to emissions but to management practices (ibid., p. 6). While the European Union lobbied for performance standards, the United States and others successfully countered that standards for pollution levels could impose trade barriers (Milliman 1995, p. 8). In the absence of emissions requirements, it is difficult to discern exactly to what ISO requirements such as "a commitment to continual improvement" refer. Does a firm commit to reducing pollution or improving management systems? Despite such ambiguities, ISO 14001 could become a way for customers in one country to compel suppliers in another to adopt more uniform environmental management systems. Indeed, some European Union countries seek to give preference to suppliers that have obtained ISO certification (ibid., p. 10).

Selection of Suppliers

It is premature to evaluate the efficacy of ISO 14000 and 14001, because companies are just beginning to implement the standards. However, a growing number of studies have attempted to consider what methods may be used to improve the environmental performance of suppliers in the electronics and computer industry (Bérubé 1992). Drawing heavily from the language of "total quality management," the literature identifies factors that encourage firms to take environmental concerns into account when selecting suppliers and during supplier partnerships (Bérubé 1992; Ellram 1990; Maxie 1994, pp. 323–329).

In a survey of 50 of the largest US computer companies, Bérubé (1992) found that, although firms are becoming more interested in integrating environmental concerns, for most companies environmental concerns are still the least important factor in the selection of a supplier. Bérubé found that large companies are much more likely than smaller ones to consider environmental issues when selecting suppliers.

Bérubé's research primarily examined what types of criteria firms use to select suppliers. A related area of research examines what types of partnerships are more successful than others in promoting manufacturing excellence. Researchers posit that partnerships that are long-term, information-intensive, and forged from a small rather than a large set of potential suppliers have a better chance of minimizing materials use and waste generation (Sarkis et al. 1995).

Bérubé's findings suggest that fabless firms are not likely to use environmental criteria when selecting suppliers. Fabless semiconductor companies are relatively small when measured in terms of staff size and sales. Recall that in 1996 SEEQ had 74 employees and slightly over $31 million in revenues. In that same year, Intel's revenues exceeded $20 billion and its worldwide workforce was near 50,000.

Another factor discouraging better supplier relationships is the extremely fluid nature of contracts, most of which are in effect for less than 5 years. SEEQ, for example, does not maintain long-term, non-cancelable contracts with its wafer suppliers (SEEQ 1997b, p. 28). Consequently, suppliers can choose to prioritize manufacturing facilities for other uses or reduce or eliminate chip deliveries to SEEQ on extremely short notice.

Foundry partnerships also may discourage environmental concerns by providing few incentives for firms to enhance manufacturing methods. Improving wafer yield requires a high degree of technical skill, the latest equipment, and close cooperation between wafer foundries and the circuit designer. However, short-term, variable partnerships and the often great geographic distance between fabless firms and overseas foundries diminishes opportunities for designers and foundries to resolve production problems (Mazurek 1994). Though partnerships among producers require information exchange between product designers and manufacturers, technology licensing and cooperative technology development may provide greater opportunities to incorporate environment concerns into product and process design. However, Angel (1994, p. 135) found that most agreements by chip design houses are for fabrication, as distinct from technology licensing or cooperative technology development.

That fabless production may fail to fit the criteria for the promotion of environmental concerns is not surprising. The current emphasis of fabless companies is on design, rather than manufacturing. Though fablessness does reduce expenditures associated with fab construction, the diminished focus on manufacturing carries risks for fabless companies as well. For example, it is not uncommon for foundries to experience manufacturing problems that result in delivery delays, or for partnerships to dissolve in less than 5 years as a result of recurrent supply or product performance problems.

Cyrix (1996, p. 15) reports: "The company's reliance on third-party manufacturers creates risks that the company will not be able to obtain capacity to meet its manufacturing requirements, will not be able to obtain products with acceptable yields, or will not have access to necessary process technologies." The experience of SEEQ illustrates what happens when a foundry partnership ends in divorce: "In the second half of '95 our revenues were adversely affected by the unexpected phase-out of one of our foundry sources." (SEEQ 1997a) The company reacted by quickly establishing ties to two additional sources.

The conditions that discourage environmental concerns among foundries may subside as third-party suppliers mature and gain the ability to invest in equipment. Indeed, TSMC, one of the largest and most established foundries, operates five fabs and plans to construct six new ones in Taiwan by 2008. The company pledges that the new fabs will deliver state-of-the-art environmental performance (TSMC 1998a). The TSMC case suggests that, as chip firms increasingly search for methods to differentiate products from the competition, voluntary initiatives such as ISO 14000 certification may become a marketing tool. TSMC (1998b) prominently touts its Taiwanese fabs as meeting ISO 14001 management standards. Similarly, National Semiconductor advertises the ISO 14001 certification of its facility in Scotland on its web site (National Semiconductor 1996).

Bérubé (1992, p. 1) finds that firms eventually may find it in their interest to incorporate environmental concerns if permitting or compliance problems lead to repeated product delays. However, such a scenario assumes that foundries are located in places with sufficient resources to scrutinize permits and conduct routine inspections. A recent analysis of

22 computer and electronics companies, including semiconductor firms based in five countries, found that, even though all the companies are based in advanced capitalist countries, half of their manufacturing and assembly facilities are in developing countries (Plazola 1997).

By most accounts, the majority of developing countries simply lack the resources to inspect and, when necessary, enforce against firms that violate standards. Taiwan, home of TSMC, is perhaps the most notorious example. The large island has serious air and water pollution and a growing number of contaminated industrial sites, owing to its rapid economic development (Arrigo et al. 1996, pp. 765–777). As environmental conditions in Taiwan have continued to deteriorate seriously, the government has recently passed a flurry of environmental laws and an enforcement agency patterned after those in the United States and Japan. Despite the proliferation of laws, Arrigo et al. report that there is little effective enforcement against polluters, primarily because of the fear that more stringent environmental protection will retard economic growth.

Singapore—another up-and-coming site of semiconductor manufacturing—implemented programs to protect the environment at an earlier stage of industrial development. Singapore introduced national air and water legislation in the early 1970s, roughly the same time as the United States, Japan, and Western Europe introduced and expanded national environmental legislation. Singapore's Clean Air Act, introduced in 1971 to control air pollution from trade and industrial premises, was amended in 1975. Singapore's water pollution control laws include the Water Pollution Control and Drainage Act of 1975 and Trade Effluent Regulations of 1976. The Poisons Act, which controls the import, transport, storage, and use of poisons and hazardous substances, applies to chemicals that have a potential for mass disaster, are highly toxic and polluting, or generate wastes that cannot be safely and adequately destroyed (Ministry of the Environment 1997). The Ministry of the Environment reports that Singapore's environmental strategies include prevention, routine enforcement, and monitoring (ibid.).

Some of Singapore's environmental programs compare favorably with those in advanced economies. According to the Ministry of the Environment, all levels of pollutants in the ambient air are within standards set by the US Environmental Protection Agency and the World Health Organization. The ministry also reports that all inland waters support aquatic life and that coastal waters meet recreational water standards (ibid.).

It is therefore likely that environmental permits and enforcement stand a better chance of promoting the environmental performance of suppliers located in countries with comparable standards and enforcement capabilities. In the context of Taiwan's regulatory climate, efforts such as ISO 14001 offer little additional assurance that a company is meeting environmental goals. At a minimum, ISO 14001 may serve as a template for companies in developing countries to put an environmental management system into place.

Although efforts such as ISO 14000 may eventually help to make environmental management practices around the world more uniform, the fluid, rapidly changing nature of foundry partnerships currently provide few incentives for chip designers and foundry manufacturers to focus on environmental concerns, particularly when foundries are located in countries that lack the resources to inspect facilities and enforce basic environmental laws. As long as foundry partnerships remain primarily an adaptive economic response to fab construction cost and increased transistor supply, initiatives such as ISO 14001 are alone insufficient to improve the management practices of foreign chip suppliers.

Interstate and Interregional Competition

The construction of new fabs away from places such as Silicon Valley carries different environmental implications than does fablessness. Even if manufacturers build fabs far from corporate headquarters, they remain responsible for the product's manufacturing and sale. Like fablessness, however, the newfound ability of chip manufacturers to build firms anywhere may reduce incentives for industry and local governments to identify and reduce environmental impacts of microchip manufacturing. Once fabs are footloose, companies can spur competition among states and among countries for the best set of locational incentives. One problem with this newfound locational freedom is that the regions most in need of fab jobs and tax revenues may have the least resources with which to address the environmental challenges associated with microchip manufacturing.

The lack of institutions to adequately track and monitor the industry's performance may actually pose unexpected environmental and economic challenges to businesses, regulators, and individuals. Consider the case of Silicon Valley. Local and regional governments there have developed regulations and ordinances tailored specifically to both toxics and the temporal requirements of chip firms. For example, Santa Clara County has provisions to minimize the potential for accidents involving hazardous gases used to make chips. The local air permitting authority also has developed provisions that approve some production changes in advance.

Similarly, environmental agencies in Arizona and Oregon have created regulations that recognize the chip industry's dynamism. For example, Arizona's Maricopa County has air permitting provisions similar to those that apply to chip companies near San Francisco that approve routine process changes in advance of actual modifications to help reduce the risk of unforeseen production delays. The Texas Department of Commerce helped to secure Samsung's decision to build a new wafer fab near Austin by offering the company assistance with environmental permits and regulations (Texas Department of Commerce 1996). As the following case studies on New Mexico and Virginia make clear, industry expansion outside established chipmaking regions carries both benefits and costs.

Intel's expansion effort in the state of New Mexico ignited both local charges of environmental racism and adverse national publicity when it elicited complaints from a small but vocal set of company shareholders. In contrast, Virginians have largely welcomed three new wafer fabs as an important new source of jobs and tax revenues. State development officials in New Mexico similarly hoped that Intel's expansion would provide the state with much-needed manufacturing jobs. However, the enormous water demands of wafer fabs are perceived by some New Mexico residents as threatening to erode an established way of life.

Intel Inside New Mexico

New Mexico, which ranks among the poorest states, is home to the Sandia National Laboratories and the Los Alamos National Laboratory. Other economic mainstays include the mining industry. The existence of Sandia and Los Alamos made New Mexico an attractive location for the state's first high-tech tenant, General Telephone and Electronics. Shortly after GTE arrived in Albuquerque in the early 1970s, other electronics firms, including Motorola, Philips Semiconductor, and Digital, followed. Intel first started operations on a 180-acre site in the Rio Grande Valley in 1980. By 1994, Intel operated three fabs near Albuquerque, and a fourth was on the drawing board (SEMI 1994, p. 3). Allied Signal and Signetics operate fabs nearby.

Some residents and some nonprofit groups that grew up with the electronics industry in New Mexico perceive Intel (dubbed by critics "Lord of the Mesa") as unresponsive to their concerns. Their mistrust was fueled by years of lawsuits against GTE for occupational health problems experienced by former employees.

Intel's problems were compounded by the public perception that New Mexico development officials had accepted Intel's terms in order to generate much-needed development dollars. To win the interstate competition for Fab 11, New Mexico agreed to finance debt stemming from Intel's plant at Rio Rancho with 30-year bonds. New Mexico also eliminated property taxes for Intel by agreeing to purchase the land where the fab would be located and to lease it back to the company. In 1992, after considering sites in California, Oregon, Arizona, New Mexico, Utah, and Texas, Intel decided to construct what would become the world's largest chip facility, Fab 11, at its Rio Rancho site.

Relative to those in richer states, environmental agencies in New Mexico have been characterized as ill-equipped to track and monitor high-tech firms. For example, New Mexico residents have historically had trouble using the TRI data because the information was not entered into a computer but was simply stored in boxes at fire stations around the state (SWOP 1995, p. 52). Another concern among those who lived near the proposed facility was that state regulators lacked the resources to verify Intel's permit applications (ibid., p. 53). To obtain permits, major pollution sources usually prepare computer models to predict possible effects of air emissions on humans and the environment. New Mexico lacked the resources to develop costly modeling studies of the potential health effects associated with concentrations of pollutants such as airborne toxics.

Fueling mistrust of Intel were concerns about environmental inequities. Intel's decision to expand production in New Mexico, where more than

50 percent of the population is Latino, followed the publication of several prominent studies that illustrate how industrial facilities and other sources of toxic pollution, including incinerators and abandoned waste sites, are concentrated in minority neighborhoods.

The topic of environmental justice first came to national attention in 1987 with the release of Toxic Wastes and Race Revisited, a report of a study conducted by the United Church of Christ's Commission for Racial Justice. The study found a higher incidence of hazardous waste facilities in minority neighborhoods (United Church of Christ 1987, p. 63). The ensuing outcry from public interest groups prompted the US EPA and the Clinton administration to adopt formal policies and procedures to discourage the redistribution of risks to minority and low-income neighborhoods. On 11 February 1994, President Clinton issued Executive Order 12898, entitled Federal Actions to Address Environmental Justice in Minority Populations and Low-Income Populations. In 1993, EPA Administrator Carol Browner had taken office with environmental justice as one of her two top priorities, the other being pollution prevention (US EPA 1994b).

Environmental justice is the focus of *Intel Inside New Mexico* (SWOP 1995), a book that criticizes Intel's New Mexico expansion effort. Subtitled *A Case Study of Environmental and Economic Injustice*, the book describes a number of blunders made by Intel in dealing with its largely Latino and Native American neighbors. The report was produced by several environmental-justice groups and sponsored in part by the Jessie Smith Noyes Foundation. The effort was coordinated by the SouthWest Organizing Project, which has tracked the semiconductor and electronics industry in New Mexico since 1989. In addition to criticizing Intel's environmental record in New Mexico, the report questions Intel's hiring practices and the tax abatements it received from the state to expand operations. The book also is highly critical of what authors refer to as "corporate welfare"—tax and land incentives offered by the state of New Mexico. As was discussed in chapter 3, states competed for Fab 11 by submitting to Intel proposals that addressed a 104-point "ideal incentive" matrix. As reported in the *San Jose Mercury News*, the "ideal" incentives included free utility service and a 50 percent price break on employees' moving expenses (Smith 1994).

Intel Inside New Mexico predicted that the incentives provided by the state of New Mexico to Intel in Fab 11's first 5 years of operation would come to $250 million (SWOP 1995, p. 45). The book also reported that Intel had received assurances from the New Mexico Environment Department that it would expedite the permitting process for the new facility (ibid., p. 49).

The authors of *Intel Inside New Mexico* contend that one of Intel's primary incentives for expanding its facilities in New Mexico was the state's relatively good air quality. Indeed, according to an article in the 20 July 1994 *San Jose Mercury News,* one of the elements in Intel's "ideal incentive" matrix was the ability to increase emissions to the air by 100 tons per year. New Mexico's air is less polluted than that in highly developed places such as Silicon Valley or Phoenix, which are designated by the EPA as failing to meet national air quality standards for key pollutants. New Mexico's relatively good air quality means that new pollution sources can emit more pollution than plants in states where ambient air quality is poorer.

In 1992, before announcing its plans to construct Fab 11, Intel received permission from New Mexico to increase air emissions from existing fabs from 140 tons to 356 tons per year (SWOP 1995, p. 52). (The fab near Phoenix that Intel completed in 1996 releases less than 160 tons of air pollutants per year.)

To some extent, the New Mexico controversy started well before Intel's announcement of its plans to build Fab 11. As was noted in chapter 2, Intel substituted ethyl 3-ethoxy-propionate (EEP) for a more volatile solvent at its existing New Mexico fabs in order to reduce the use of volatile organic compounds that contribute to urban smog. In addition to EEP, other chemicals emitted from Intel's older New Mexico facility were acetone and isopropyl alcohol. Intel's use of EEP caused an inadvertent odor problem at four Intel fabrication facilities, including those in New Mexico. In New Mexico, however, residents complained of the new smell. With the GTE case still fresh in the minds of many, neighbors sought assurances from Intel that the fumes they were smelling were indeed safe to breathe. Intel tried to reassure residents that the fumes were from EEP, which can cause drowsiness, and not a more hazardous substance, such as acetone, which can affect the central nervous system. Despite Intel's assurances, a number of neighbors continued to worry that the smell was from acetone.

Though it was never established that the smells were from acetone or that neighbors were exposed in sufficient quantities to experience adverse health effects, Intel's perceived lack of responsiveness to local concerns dampened its relations with the community and with some environmental regulators. The newspaper *Corrales Comment* reported: "For some time now, state regulators and other local government officials have complained of Intel's arrogance and presumptiveness." (Radford 1993)

Intel completed Fab 11 in 1995. Since then, Intel's greatest perceived offense in New Mexico involves its need for water. According to the 23 March 1996 *Corrales Comment,* Intel's New Mexico facilities used between 2 million and 3 million gallons of water per day before Fab 11 was built. After Fab 11 came on line, Intel's daily water use increased to between 4 million and 5 million gallons, and the 10 January 1996 *Albuquerque Journal* suggested that this might increase to 6 million gallons per day—roughly 5 percent of the amount used in Albuquerque.

Intel reports that it has experimented with wafer cleaning machinery that would require less ultra-pure, de-ionized water. In New Mexico, Intel conducted a pilot project to recycle wastewater for use in plant cooling. Intel found that such diversions could reduce the demand for fresh water by 500,000 gallons per day (Intel 1995, p. 9). Such advances help to reduce but do not eliminate semiconductor manufacturers' needs for a resource that in New Mexico and other parts of the desert Southwest is increasingly scarce.

In 1997, some of the same groups that published *Intel Inside New Mexico* released *Sacred Waters: Life-Blood of Mother Earth,*[2] a book that charges chip makers with draining the Southwest's scarce water supplies and contaminating groundwater. *Sacred Waters* critically examines water use and disposal in four Western high-tech hubs, including New Mexico.

Sacred Waters pulls together years of nonprofit-group research to paint a bleak portrait of the region's future. Industry representatives counter that the book contains some inaccuracies and fails to document advances

2. *Sacred Waters* was published by the Electronics Industry Good Neighbor Campaign, a collaborative effort of the Southwest Network for Environmental and Economic Justice and the Campaign for Responsible Technology.

made by companies such as Intel in recent years to reduce water use and discharge. One Intel Arizona plant manager told the *Arizona Republic*: "We're stationary targets to some extent, we're visible, we're manufacturers and we're easy to attack." (Fehr-Snyder 1997, p. 2) In view of the constantly shifting mix of chemicals required for wafer manufacturing, industry says that "closed loop" recycling methods are an obvious goal. However, companies maintain that the goal remains elusive, in part because the industry's constantly shifting mix of chemicals makes it is difficult to identify and adequately remove what is in the waste stream at any one given moment in time.

Sacred Waters throws into question the long-term ecological viability of wafer fabs in the desert Southwest. New Mexico is a case in point. Intel is not the only entity in the desert Southwest that demands water. Since 1950 New Mexico's population has tripled to just over 1.5 million, according to the 1990 census. The state's continued residential and industrial growth has tapped limited water supplies. Before 1900, the Rio Grande provided residents of the Albuquerque Basin with sufficient water. However, competing water demands from ranching, agricultural, and residential use have since led the state to over-appropriate river supplies. Today, residents of Albuquerque draw all their drinking water from underground aquifers that connect to the Rio Grande through recharge.

To meet its water needs, Intel has embarked on a process that has caused some observers to seriously question what the continued expansion of microchip manufacturing may portend for New Mexico and other southwestern states. Intel has applied to the New Mexico State Engineer for permits to drill three wells into an underground aquifer that is ultimately linked to the Rio Grande. Considering Intel's water needs, the state has determined that the company's pumping of groundwater could eventually affect the volume and flow rate of the Rio Grande (EIGNC 1997, p. 84). To offset potential effects on the Rio Grande from groundwater pumping, the state ordered Intel to purchase and retire rights to 2207 acre-feet of water from the Rio Grande. Upon purchase, the rights would be transferred as a credit to offset water withdrawn through pumping.

Intel did not anticipate how its overtures to purchase water rights from indigenous family farmers would be received. The film *Chinatown* illustrates the lengths to which some speculators went to bring water to the desert. The film, starring Jack Nicholson, is loosely patterned after the story of William Mulholland and the downtown power brokers of Los Angeles who made their fortunes—and turned a dusty little town into sprawling metropolis—by, in effect, robbing ranchers of their water in the Owens Valley, hundreds of miles north of the city.

Fictionalized accounts of Western water wars such as that portrayed in *Chinatown* depict a stance toward water allocation that is relatively recent in origin. According to market-based perspectives, water should be allocated to the production processes that have the highest per-unit production value. In other words, if one microchip is valued at $100 and one bushel of corn at $50, then water should be allocated to microchip manufacturing. The water policies of the American West have sought to divert water to its highest use. However, that is not consistent with the view held by the Pueblos, who have lived in North Central New Mexico for hundreds of years. For the Pueblos, water has historically represented a communal asset—a perception that persists in some parts of the state to this day.

By one account, the Pueblos and the Hispanic settlers organized economic and social life around an elaborate earthen irrigation system known as *acequias* (Rivera 1996, pp. 11–15). In most villages, local government positions are structured around the acequia, with posts such as ditch commissioners, ditch bosses, and water system members.

According to *Sacred Waters*, life from farm to farm is stitched together through the irrigation system. For example, it is understood that it is each member's duty to keep the acequia clean in order to maintain water flow. Members voluntarily maintain the ditches that traverse their land because it is understood that if one part of the system fails neighbors downstream will suffer (EIGNC 1997, p. 85).

The communal acequia system stands in stark contrast to the economic, engineering, and legal principles that govern water management in the desert Southwest. Indeed, the idea that water represents a commodity is itself alien. The locals have this expression: "La tierra es la madre, y el agua es su sangre." (The land is our mother, and the water is her blood.) Thus, when Intel sought to purchase water rights to offset pumping, the company's search was perceived by some as a potential threat not only to water supplies but also to social traditions set down over the centuries (EIGNC 1997, p. 85).

Intel's search for water has come to symbolize a clash between New Mexico's past and its potential high-tech future. As portrayed in *Intel Inside New Mexico,* the company's proposed purchase of rights to scarce water would accelerate the transformation of surrounding areas from small farms to three-acre estates owned by Intel engineers. According to the text, "the gentrification of Corrales and its neighbors will only accelerate . . . if the water right is severed and all you can grow are subdivisions and fast food joints" (SWOP 1995, p. 56).

Water use is the most volatile and complex issue surrounding Intel in New Mexico. However, some have also questioned the company's wastewater treatment plans. Intel agreed to share the cost of upgrading the local treatment facility with Albuquerque to handle the flows from expansion. Yet opponents alleged that taxpayers would underwrite most of the cost of sewage line upgrades and expansion (SWOP 1995, p. 61). *Intel Inside New Mexico's* authors also are critical of the company's decision to pump wastewater some 20 miles from the plant to a treatment facility located largely within a Latino neighborhood.

In 1994, Intel launched a campaign to improve relations with its neighbors in New Mexico. As part of the campaign, Intel representatives met with local groups such as the Corrales Residents for Clean Air and Water. However, the move was apparently too late. By the time that Intel launched its community relations campaign, its problems with its neighbors in New Mexico had come to the attention of the Jessie Smith Noyes Foundation. Among the New York foundation's holdings were 1000 shares of highly profitable Intel stock. In 1994, in response to the New Mexico controversy, the foundation filed a resolution saying that Intel selected "environmentally risky" sites for its operations. According to the *Chronicle of Philanthropy,* the resolution received 5 percent of stockholders' votes (Greene 1996, p. 25). An additional 8 percent abstained. The *Chronicle* observed that, although it is not unusual for foundations to promote various causes by funding grant proposals, the Jessie Smith Noyes Foundation's decision to target its own stock holdings was largely unprecedented.

Local activists and those who signed the shareholder resolution say that the adverse publicity helped persuade Intel to revise its corporate environmental health and safety policy, but Intel denies the claim (Greene 1996, p. 25). The company did offer to meet periodically with New Mexico activist groups such as the SouthWest Organizing Project.

No Battle at Bull Run

Intel ultimately picked New Mexico over other states as the site for Fab 11 because it already had chip factories there. However, the advent of advanced design and manufacturing facilities now makes it possible for companies to build new production facilities far from existing fabs. Economic development planners in Virginia attracted three fiercely recruited semiconductor wafer fabs, which they hope will anchor the "Silicon Dominion" (VEDP 1997, p. 8).[3] Under the banner of "the bottom line state," Virginia's recruitment efforts in business magazines focused on the state's access to markets, its affordability, and its quality of life. Northern Virginia already has the largest concentration of computer communication companies in the country. Fairfax County, a suburb of Washington, is home to 1600 technology companies. Many of these companies create applications for the Internet, which was developed by the Pentagon. Despite the growing presence of communication companies, microchip manufacturing is relatively new to Thomas Jefferson's home state.

In rural Goochland County, near Richmond, sewage, water, and electrical hookups are in place for Motorola's $3 billion West Creek facility, which will be the largest of the state's three new semiconductor plants and Motorola's largest single capital investment ever. In 1997, in Henrico County, Motorola and Siemens in set a new industry standard by going from greenfield development site to equipment installation in just 11 months. The White Oak semiconductor facility, located on 210 acres near Richmond International Airport, began producing memory chips in 1998.

Virginia's first new fab, in Manassas, makes 64-megabit DRAM chips.[4] In August of 1997, the first chips rolled off the $1.7 billion Dominion Semiconductor's first production line. White Oak, the Motorola-Siemens

3. "Old Dominion" is a traditional nickname for Virginia.
4. The 64-megabit technology was jointly developed by IBM, Siemens, and Toshiba.

joint venture outside Richmond, also manufactures limited quantities of 64-megabit DRAM chips; however, it will eventually make the more lucrative SRAM chips.

IBM and Toshiba broke ground on their Northern Virginia joint venture in November of 1995, four months after the site was officially selected. One line is currently in operation inside the 600,000-square-foot fab, and planning of a second fab of similar size is already underway. The master plan for the 120-acre site can accommodate up to three fabs, which may employ up to 4000 people.

In contrast to Motorola's plants outside Richmond, the Dominion site in Northern Virginia has long housed electronics and limited microchip manufacturing. Dominion's new 64-megabit fab sits adjacent to a former IBM facility that closed its doors in 1992 due to failing markets and rapidly advancing technology. Dominion converted the former IBM fab to serve as office space for the new plant. The presence of the existing facility, which once employed 5000 workers, was a factor in Dominion's decision to locate the $1.7 billion facility in the city of Manassas.

To accommodate the move, state and city officials offered tax incentives, mostly in the form of reduced taxes on equipment. Recall from chapter 1 that equipment represents roughly 75 percent of a fab's total cost. Planners in neighboring Prince William County also have strongly supported the Dominion plant, which they hope will help draw similar chipmaking facilities. If Dominion builds all three fabs, incentives from the state of Virginia will add up to about $48 million over 20 years (Behr 1997). Of this total, about $38.4 million is in the form of performance-based grants paid over 5–10 years. Job tax credits come to $3.8 million. Workforce training and education account for the remainder. The city of Manassas pledged to provide around $100 million, mostly in the form equipment tax relief over 10 years, according to a report in the 16 June 1996 *Washington Post*. State and local officials estimate that the facility could generate up to $477.5 million in direct and indirect taxes.

IBM has maintained a presence in Manassas for several decades. Motorola led the recent charge into Virginia in 1995, when it announced plans to build the West Creek facility in Goochland County. Eastern Virginia has developed a base in defense and shipbuilding, whereas Richmond, considered to be in the central portion of the state, has historically been tied to tobacco. However, a number of factors, including shrinking defense dollars, rail consolidation, and the shifting public sentiment against tobacco companies provide state development officials with strong incentives to lure a new industry.

The vagaries of chip markets make it unclear at this point just to what extent microchip manufacturing will take hold in Virginia. The IBM-Toshiba fab in Manassas started production in August of 1997. Around the same time, Motorola placed the construction plans for its $3 billion West Creek microprocessor facility on indefinite hold as a result of Apple's failing fortunes. (According to Motorola's original plans, West Creek was to make PowerPC microprocessors for Macintosh computers.)

In late 1997, Motorola revised its plans for the West Creek site. Instead of microprocessors, Motorola now plans to make components for pagers and facsimile machines. To accommodate the change, the plant will be twice the size of the one originally planned. When complete, the five-building, 1.5-million-square-foot complex will initially create 2500 jobs. Motorola officials said that the plant ultimately could employ up to 5000 people, according to the 4 December 1997 *Washington Post*. Diversifying its manufacturing portfolio may reduce Motorola's vulnerability to market swings. Still, volatility in any one segment of the industry may still cause Motorola to put its plans for the West Creek site on hold. Market shifts caused Motorola to modify plans for its second facility near Richmond. White Oak Semiconductor started producing limited DRAM supplies in 1998. However, the site will be retooled somewhat because Motorola in 1997 decided to phase out DRAM memory chip manufacturing. Finally, although Dominion Semiconductor hopes to eventually operate three wafer fabs at its site in Manassas, Dominion's expansion plans ultimately will be determined by demand in the highly volatile DRAM market.

State officials must be applauded for luring chip fabs to a region with no established cluster of semiconductor companies. However, some observers question whether environmental concerns were part of the state's economic development strategy. Virginia's successful bid to lure wafer fabs coincides with a string of environmental enforcement incidents under the administration of Republican Governor George Allen. Under his administration, the Virginia Department of Environmental Quality was reorganized. Some VDEQ employees say that "reorganization" was a euphemism for curbing their monitoring and enforcement capabilities. In 1997, US EPA Administrator Carol Browner took the state to task, charging that Virginia suffers from a "lack of leadership" and displays a "disregard for the requirements of federal laws" (Cushman 1997). The state's enforcement record against polluters is so weak that the EPA in July of 1997 threatened to take the task of managing air, water, and waste problems away from VDEQ and manage the state's environmental affairs from nearby Washington (Hsu 1997).

There is no evidence to suggest that state economic development officials relaxed environmental laws to lure the three chip makers. However, as the experience of Dominion Semiconductor case illustrates, some local officials failed to consider in advance of luring the company one of the most obvious environmental requirements of these mammoth manufacturing facilities: sufficient capacity for the disposal of millions of gallons of wastewater per day.

To commence operation on the single production line in August of 1997, Dominion obtained permits to dispose of 2.28 million gallons of water daily through the Upper Occoquan Sewage Authority. The regional authority is run by representatives from Fairfax, Fauquier, Loudoun, and Prince William Counties and from the cities of Manassas and Manassas Park. In 1996, concerned about wastewater, representatives from these jurisdictions formed a regional committee to evaluate the impacts of the chip plant. After an initial evaluation of the Dominion plant, which straddles the border of the city of Manassas and Prince William County, supervisors from the neighboring county of Fairfax expressed concern about sodium levels from the plant's wastewater discharge. To neutralize acids in wastewater, plants such as Dominion typically treat wastewater with sodium chloride or salt before releasing it to a public sewer system. After treatment, some of the water would be returned to the Occoquan Reservoir, which provides drinking water to about 500,000 residents in parts of Prince William and Fairfax counties. According to a 24-page evaluation prepared for Fairfax County Board of Supervisors, residual sodium levels in the wastewater would be sufficiently high to pose potential risks to some sensitive segments of the population, such as pregnant women and those who suffer from heart trouble or high blood pressure (White 1996).

Initially, Dominion and the city of Manassas had proposed constructing a costly treatment plant that would use a reverse osmosis filtration process to filter wastewater well enough to send most of it to Lake Manassas, which supplies drinking water for the city. However, regional officials rejected the plan because they maintained that such a plant would create a residue, known as brine, that contains very high levels of sodium. Chip manufacturers in arid regions typically discharge salt water to ponds where it evaporates, leaving dry sodium behind. Such a solution is less feasible in Northern Virginia's relatively wet climate.

Another option was for Dominion to modify the manufacturing process to require less sodium. Yet another was to develop better methods for sodium abatement. A Fairfax County Supervisor was quoted as follows in the 24 August 1996 *Washington Post*: "This may be very expensive for the company to do, but that to me is the cost of doing business." Ultimately, Dominion Semiconductor opted to modify manufacturing processes to reduce sodium levels discharged to drinking water supplies.

Although manufacturing modifications allowed the Dominion plant to secure an initial operating permit for a single wafer line, the Upper Occoquan Sewage Authority will ultimately be unable to handle the increased volume of wastewater from the plant's proposed expansion. To allow for expansion, Dominion Semiconductor has applied to the VDEQ for permission to build a 26-mile pipeline through Prince William County that would funnel as much as 8 million gallons each day of treated wastewater into Powells Creek and Quantico Creek, tributaries of the Potomac River (Blum 1997). A nonprofit conservation group, American Rivers, lists the Potomac as the nation's seventh most endangered river. On the mend from severe pollution problems, the Potomac flows into the Chesapeake Bay, one of the nation's richest estuaries. The Chesapeake Bay has been damaged in recent decades by overfishing and by nutrient runoff from farming and development.

Prince William County appointed a citizens' advisory committee to study potential impacts of the proposed pipeline in 1997. The committee concluded that "the Dominion Semiconductor wastewater discharge will not have a discernible water quality impact on Quantico Creek, the Potomac River, or the Chesapeake Bay" (CACIWSS 1997, p. 19). However, the committee also concluded that Powell's Creek, one of the two creeks proposed to handle the flows, is not an acceptable location for industrial wastewater discharge. According to a report issued by the citizens' committee in November of 1997, wastewater flows into that tributary would "alter the existing environmental resources of the creek" (ibid., p. 19). The committee also recommended that stringent erosion and sedimentation controls be developed to prevent runoff that would result from construction of the underground pipeline, which would be roughly 20–24 inches in diameter. The committee also recommended that the pipeline contain equipment for monitoring or controlling spills and leaks by routinely performing failure analysis or testing pressure. Finally, to the maximum extent feasible, the report recommends that the plant develop strategies for recycling and reusing water.

The Dominion case is distinguished from other major proposed developments in Manassas and surrounding areas by the lack of opposition the fab has stirred. Best known to most Americans as the site where two great armies rocked the foundation of a struggling nation in the summers of 1861 and 1862, Manassas, as well as neighboring Prince William County, in recent years has witnessed pitched battles between historic preservationists and real estate developers. Since the mid 1980s Manassas has seen many of its farms transformed into subdivisions served by strip malls.

Located about 30 miles west of Washington, Manassas is near a historic battlefield named after a small creek or "run." The Battle of Bull Run helped to determine the fate of the young republic. In his highly acclaimed 1991 book *Edge City*, Joel Garreau describes how during the late 1980s a hodgepodge of preservationists and history buffs successfully rallied to fight a proposal for a shopping mall and got the site designated a National Battlefield Park. In 1994, not long after the "Third Battle of Bull Run," Virginians in nearby Prince William County blocked the Disney Corporation's plans to build a US-history theme park.

In contrast to the pitched battles waged by environmentalists against developers and Disney, the Dominion plant has managed to strike a balance between the region's past and its possible future, as the plant's name suggests. For example, inside the perimeter of the 120-acre Dominion site is a Civil War cemetery whose graves are still tended by soldiers' descendants. Just across the street from the site stands a ramshackle set of red farm buildings—a testament to the former mode of production that characterized the region.

To address potential environmental issues from the onset, the city of Manassas formed a citizens' committee in 1996 to study the air, water, and waste impacts of the Dominion plant. According to one of the few neighbors who continues to attend city council meetings and press plant officials for information about its environmental effects, the official citizens' committee lost momentum and disbanded after about 8 months (Wood 1997). In contrast to the activists who sought to stop development near Bull Run and at Disney, however, the Dominion facility was built on a site that has long served industrial uses as an IBM facility.

IBM's long occupancy of the site probably helped to minimize potential resistance to development. However, earlier manufacturing operations there also helped contribute to groundwater contamination. Consistent with the industry's groundwater problems in Silicon Valley and Phoenix, IBM is currently filtering groundwater at the Dominion site to clean up an underground perchloroethylene plume. The popular dry cleaning solvent entered the groundwater when the former IBM facility was still in operation. According to Dominion officials, the spill was caused by a contractor who improperly drained hoses at the site after filling supply tanks. A nearby dry cleaning plant also is suspected of having contributed to the contamination. Nonetheless, IBM agreed to assume full responsibility for the spill as part of what is known under the Resource Conservation and Recovery Act as a "corrective action." Dominion says IBM spends about $1 million a year to pump and filter the groundwater, which is not part of public drinking water supplies.

Since the 1980s, chip companies have taken a number of steps to prevent soil and groundwater contamination. Many of these steps are in evidence at Dominion. For example, whereas it was once common practice for chip makers to store chemicals on site, today, most major fabs—including Dominion's—have moved to "just in time" systems, trucking in supplies as they are needed. And whereas companies once concealed supply hoses, tanks, and ducts underground, tanks and wastewater pipes at the Dominion plant are now elevated on tracks above ground to promote early leak detection.

Despite such improvements, Dominion—like most other chip companies—still uses potentially hazardous and toxic chemicals, including arsine and phosphene. Furthermore, some of the chemicals are trucked through the local community. Plant personnel have worked with the local fire department, hospital, and hazardous materials team to develop a local emergency response plan, as is required under the Emergency Planning and Community Right-to-Know Act.

Perhaps owing to their lack of familiarity with toxic and hazardous chemicals used in microchip manufacturing, most local residents have expressed more concerns about economic development and land use issues associated with the facility than about potential chemical risks. Most people who turned up at city council meetings after IBM and Toshiba selected the site question city officials' decision to offer incentives to a company that might one day leave. Still smarting from the jobs lost when IBM closed its original facility in 1992, some Manassas residents worried that their hard-earned tax dollars would lure a facility that ultimately would suffer the same fate as the original plant. Since chips started to roll off the Dominion plant's first line in August of 1997, however, such fears have largely subsided. Other neighbors met with plant officials to express their desire to keep land slated for the plant's wastewater holding pond as open space.

Dominion's policy has been to meet with concerned residents, supply them with information, and invite them to the site. By most accounts, people are satisfied with Dominion's policy. However, one or two neighbors continue to question whether it is safe to live near a wafer fab. One of them is Maureen Wood. A former member of the local citizens' committee that formed to study environmental issues, Wood has lingering questions about the health and environmental effects of chemicals used to fabricate wafers. Her concerns intensified after NBC, in late 1997, aired its *Dateline* segment about IBM workers who allege that chemicals they used caused cancer. In order to learn more about the potential impacts of microchip manufacturing, Wood has been required to contact organizations largely outside the region for information.

Though Northern Virginia and neighboring Washington house some of the most prominent environmental organizations in the country, until now there has been no need to develop regional institutions focused on microchip manufacturing. Instead, most local environmental groups have trained their efforts and resource dollars on existing issues, such as land conservation. In order to learn more about the potential impacts of the plant, Wood was therefore required to spend her own time and money to obtain information from industry watchdog groups in Massachusetts and California (Wood 1997).

Dominion representatives acknowledge that more people will likely raise similar questions about the risks of chip plants as information about the industry's history in places such as Silicon Valley becomes more widespread. In view of the enormous investment IBM and Toshiba have made to locate the facility in Manassas, company officials say that they are eager to demonstrate that their plant will be a good neighbor. Doing so, they say, requires greater industry openness and responsiveness to public concerns. According to a Dominion spokesman, the only information that they refuse to release to the public are salary figures, yield rates, and other sensitive data that might be used by competitors. In addition to greater openness, Dominion representatives say that one of the most important things that state and local governments can do to improve public confidence in the industry is to develop strong environmental monitoring and enforcement policies. Dominion's community and government relations representative has said: "There must be a perception that environmental protection has some parity with economic development goals." (Holcomb 1997)

Despite assurances from Dominion officials, the new plant has experienced several environmental mishaps. About 8 months after the plant commenced operation, the liner in one of the wastewater holding ponds cracked, allowing contaminated water to leak (Wood 1998). Several months later, 1200 gallons of a caustic alkaline chemical spilled from a leaky pipe into a nearby creek, according to a report in the 12 May *Washington Post*. Dominion workers blocked the creek—a tributary of a stream called Cannon Branch—and drained it until the water's alkalinity returned to normal. Dominion officials told the *Post* that the spill posed no health risks to humans. However, it killed an unspecified number of small fish.

Dissenting Views

Like Dominion's effort in Virginia, Intel's New Mexico expansion effort was the result of an interstate competition for fab dollars. In both cases, the construction of a new fab occurred not at a greenfield location but at a site containing preexisting semiconductor manufacturing facilities. Though both firms elected to expand facilities in states that appeared to hold out the best mix of locational incentives, Intel encountered intense public opposition—opposition that ultimately resulted in adverse national publicity.

Why did Intel attract so much opposition and Dominion so little? One factor may be timing. Intel's experience sounded a wake-up call for the industry in regard to the way it does business with local government and neighbors. Since the publication of *Intel Inside New Mexico,* Intel has taken a number of steps to promote community relations. One such effort forms the focus of chapter 6.

In contrast to Virginia, Intel's expansion effort in New Mexico took place in an established chip and high-tech manufacturing region that had witnessed an unfortunate incident at a now-closed facility owned by General Telephone and Electric. Approximately 225 former GTE workers have been involved in lawsuits against the company stemming from suspected cancers and other illnesses which they maintain stemmed from working at the plant (SWOP 1995, p. 23). Owing in part to the GTE incident, Albuquerque supports a number of well-established environmental and environmental-justice organizations that focus specifically on the downsides of high-tech manufacturing.

As the title of the *Intel Inside New Mexico* tome suggests, the company's experience there also was perceived by some as a case in which a company sought to redistribute risk away from states with more affluent, Anglo populations to a relatively poor state that is over half Latino and Native American. Though Latinos and Asians make up an increasing percentage of Northern Virginia's population, combined they account for only about 13 percent of the total, and African-Americans for another 26 percent, according to the US Census Bureau. Environmental-justice organizations in New Mexico are concerned that other chip and electronics firms will perceive New Mexico as a pollution haven and seek to locate manufacturing operations there. Public mistrust of Intel in New Mexico also was fueled by the perception that state and local governments possess inadequate resources to evaluate Intel's proposals and protect the health of people and the environment.

To some degree, the Bull Run, Disney, and Intel battles illustrate intense public fears about the social and economic transformations development will bring about. The Northern Virginians who blocked a developer's bid to build a shopping mall on the site of a historic battlefield have something in common with the New Mexico residents who worry that high-tech companies will replace a centuries-old form of social organization with fast food joints and expensive houses.

Social anxiety, as urban sociologists like to show, is maladjustment to change. As the millennium draws to a close, perhaps nothing promises to transform how people work and play more than the microchip. Unlike Intel in New Mexico, Dominion Semiconductor has managed to strike a balance between the region's past and future. In contrast to New Mexico, residential and commercial development—not microchip manufacturing—converted Manassas and Prince William counties from agriculture to their largely exurban uses. Dominion provides exurbanites with a relatively high-paying alternative to work in retail and service sectors. However, the Disney and Bull Run experiences show that public sentiment could change if planners bring more fabs, and hence more houses and strip malls, to Northern Virginia.

Intel's experience in New Mexico serves as a cautionary tale for chip makers who seek to expand operations. As the controversy surrounding Intel's search for water shows, other factors beyond tax incentives and regulatory streamlining must go into a company's locational calculus. Intel officials failed to consider the degree to which the acequia principles that guide land use and governance in North Central New Mexico are distinct from values that drive corporate decisions in Silicon Valley. The notion that water should be used where its use value is highest—in this case, wafer fabs instead of subsistence farms—struck a number of people in New Mexico as foreign, if not somehow perverse. To some, the acequias and attendant customs and values seem a far more rational way of organizing life than the stark image of wafer fabs and fast food establishments that microchip manufacture portends. Thus, Intel's quest for water came to symbolize much more than a routine business transaction. For some, New Mexico's silicon development strategy threatens to destroy a system of social organization rooted deeply in the region's past.

Summary

Changes in how and where US semiconductor firms make chips outstrip the present ability of government institutions to track and monitor potential threats to humans and the environment. Mechanisms that could help to improve information about the industry, such as the Toxics Release Inventory, are increasingly inadequate, owing to complicated reconfigurations in how and where chips are produced.

New foreign chipmaking regions, such as Taiwan's HsinChu, recently have enacted environmental laws with standards that are comparable to those on the books in the United States or Japan. These countries, however, often lack the resources and the political will to conduct inspections and enforcement.

Voluntary initiatives to improve environmental performance among suppliers, including supply chain management and ISO 14000, appear unlikely to improve the performance of foreign subsidiaries and foundries in the near term. However, if production delays or product defects begin to outweigh the benefits of short-term supplier relationships, it may be possible to use supply chain management techniques to improve how the market segment manages environmental problems.

The lack of an established industry base and supporting institutions also may offset some of the benefits associated with established chipmaking regions, as the case of Dominion aptly illustrates. A policy of openness can help to minimize potential opposition. However, for plant neighbors who seek independent technical information about health and safety effects of living near fabs, the lack of organized industry watchdog groups also may impose higher costs. Similarly, the perception among some New Mexico residents and interest groups that local regulators lacked sufficient resources and muscle to monitor and assess the environmental impact of Intel's expansion contributed to the firm's troubles in New Mexico—and, potentially, on Wall Street.

Finally, though the current explosion in new fab construction may deliver short-term economic benefits, it is questionable whether states should pin their long-term development hopes on an industry in which Moore's Law may one day no longer hold. The questionable future of chipmaking methods based on photolithography ultimately raises the issue of what will happen to the Dominion and Intel facilities after 20 years—an issue that the industry, the planners, and the environmental groups have largely failed to tackle.

The issue is perhaps most poignant in Virginia, which, unlike New Mexico, has not experienced the downsides of high-tech manufacturing. For years, Virginia has billed itself as a state that "is for lovers." It is hard to imagine how the bucolic appeal of Jefferson's home state will mesh with development strategies based on the notion that "only the paranoid survive." Although three wafer fabs will provide much-needed jobs and revenues, it is not clear whether Virginians will support plans to pattern parts of the state after Silicon Valley.

5
Cleaner, Cheaper, Smarter[1]

Intel's New Mexico expansion effort illustrates the growing gap in perception among industry, public interest groups, and individuals about the human and environmental health effects associated with microchip manufacturing. Industry claims that its processes are clean and merit fewer cumbersome regulatory restrictions. Conversely, some public interest groups and concerned residents want greater assurances that fabs make good neighbors.

In its characteristic spirit of conciliation, the Clinton administration sought to bring both sides closer together by launching the Common Sense Initiative (CSI) and Project XL, two high-profile initiatives designed to improve the effectiveness and efficiency of environmental regulations. This chapter describes the genesis of both initiatives and examines the promise and pitfalls of CSI.

Of all the potential sectors that CSI could target, none appears better suited than computers and electronics. The nascent, dynamic manufacturing form appears to collide squarely with the cautious, deliberative system of environmental law and administration that grew up during roughly the same time. Instead of tackling tough issues of dynamism head on, CSI has focused on less controversial and economically disruptive remedies to a few environmental problems associated with electronics and computer manufacturing. Though the Environmental Protection Agency designed CSI to achieve reform by bringing together groups that commonly act as adversaries, it is not clear whether the initiative has served to successfully bridge the growing gap that divides regulators,

1. This is the motto of the US EPA's Common Sense Initiative.

public interest groups, and industry. To better understand the strengths
and weaknesses of CSI, it is instructive to first examine some of the argu-
ments as to why traditional regulations are deemed increasingly
inappropriate.

Dinosaur Regulations?

Even in industries where production methods and physical plants are
more stable than they are in microchip manufacturing (for example,
petroleum refining), no two facilities are identical. Yet individual laws for
air, water, waste, and toxics mostly treat facilities and industries as uniform. In general, environmental laws fail to focus on sectors or facilities;
instead they direct regulators to determine what concentrations of a pollutant may harm humans and the environment. Some exceptions include
the New Source Review provisions under the Clean Air Act Amendments
of 1990, which set standards on an industry-wide basis. For pollution
concentration standards, agencies usually develop technological controls,
permits, and reporting requirements for pollution sources.

To some degree, the present system is a success. Pollution levels for air,
water, and, to a lesser extent, waste and toxics have declined since Congress developed most of the major federal pollution laws in the late 1960s
(Davies and Mazurek 1997). Uniform national laws also benefit people
and companies in other ways. For example, they help firms with operations in many states to meet one set of requirements rather than 50 different requirements. The laws also ensure that people who travel from one
state to the next are afforded similar health protection. Finally, current
laws help to discourage "smokestack chasing," a practice whereby states
compete for polluting industries by relaxing regulatory requirements.

Yet even at petroleum refineries, uniform laws sometimes increase costs
without necessarily controlling the chief causes of pollution, as the US
Environmental Protection Agency and Amoco learned at a refinery in
Yorktown, Virginia. In a novel exercise, industry, regulators, and public
experts jointly studied the production characteristics of the plant and
found a cheaper way to control pollution more than regulations required.
The group discovered that the refinery could control air pollution in a
more cost-effective manner by capturing vapors from fuel pumped to
barges instead of by restricting emissions at the refinery. The group fur-

thermore found that regulations actually prevented Amoco from selecting the least costly method to control volatile organic compounds (Amoco and US EPA 1992).

Cleaner, Cheaper, Smarter

By identifying and, if necessary, removing barriers such as those identified at the Yorktown refinery, the EPA developed the Common Sense Initiative as a way to make regulations "cleaner, cheaper, and smarter." Unlike the Amoco Yorktown project, which focused on a single refinery, the EPA has aimed CSI at about eight industries, organized into what the agency refers to as "sectors." The EPA's six sectors are computers and electronics, auto manufacturing, iron and steel, metal finishing and plating, petroleum refining, and printing. The EPA's terminology is somewhat misleading because it departs from Standard Industrial Code designations. In contrast to the EPA's computer and electronics sector, the SIC codes distinguish between the computer industry and the broader electronics sector (SIC 36). SIC 36 includes semiconductor makers, producers of printed wiring boards, and some segments of the computer industry. Other segments of the computer industry fall under SIC 35.

For each of its six sector groups, the EPA convenes participants traditionally thought of as adversaries: industry, local and state agencies, and representatives from environmental, labor, and environmental-justice organizations.

Before 1994, when the EPA implemented CSI, the initiative's architects hoped that it would serve as a vehicle for participants to identify and, if necessary, eliminate conflicting or contradictory regulations. CSI's designers also hoped that the initiative would help to make preventing pollution, rather controlling it, a standard business practice. Finally, they envisioned CSI as a way to give industry incentives and flexibility to develop innovative technologies that would meet or exceed environmental standards. As the experience of the ongoing initiative suggests, some of the CSI's early goals now appear overoptimistic.

Despite its shortcomings, CSI's focus on electronics and computer companies remains promising: the initiative is a way for the EPA to determine whether environmental regulations unduly hinder fast-moving microchip, computer, and software companies.

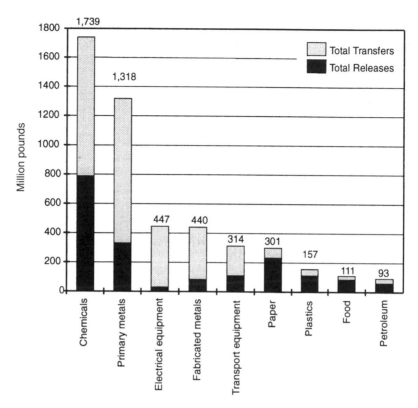

Figure 5.1
TRI releases and transfers in selected sectors, 1995. Source: US EPA 1997b, pp. 30–31.

As envisioned, the initiative appears exceptionally well suited to examine whether current regulations effectively target the electronics sector (SIC 36). Despite the sector's clean appearance, the EPA reported in 1994 that toxic releases and transfers off site for the computer and electronics sector were the third highest in the United States for manufacturing sectors required to report that year. Figure 5.1 presents a partial listing of industrial sectors required to report and TRI totals for 1995. Among all manufacturers required to report, TRI totals were highest for the chemical sector, followed by primary metals (US EPA 1997b, p. 34). TRI releases from the electronics sector in 1995 were slightly more than 30 million pounds (ibid., p. 30). Releases from chemicals were over 787 million pounds, and those from the primary metals sector were about 331 million pounds. Though releases from the electronics sector were relatively low, toxic transfers off site for the sector came to over 416 million pounds (ibid., p. 31). Total transfers off site for the chemical sector in 1995 were about 951 million pounds, and about 986 million pounds for the primary metals sector.

In addition to high TRI transfers, continuous innovation in chip manufacturing also contributes to solid and hazardous waste problems in the electronics and computer sector. Newer, faster chips make hardware obsolete. In contrast to current environmental laws, CSI's sectoral focus has the potential to reduce toxic use by encompassing firms all along the product supply chain—from inputs such as semiconductors to final products, including personal computers and cellular phones. CSI's sectoral focus also has the potential to address environmental problems associated with US manufacturers' growing reliance on offshore semiconductor foundries and electronics assembly operations.

Because the Common Sense Initiative has the potential to target simultaneously suppliers and manufacturers of intermediary and final goods, it represents an opportunity to develop data to better guide management of constantly shifting sets of process chemicals used by chip and computer firms.

In practice, CSI's aims and achievements have been modest. The initiative's broad focus is constrained primarily by the lack of legal mechanisms to make sectoral projects operational. In other words, CSI participants have largely avoided projects that would go beyond the current regulatory system of separate air, water, and land laws precisely because there exists no statutory authority to do so (Davies and Mazurek 1996, pp. 19–30). Instead, most of CSI's six sectors have proposed pilot projects that represent incremental improvements possible under the extant regulatory system. These projects are largely designed to reduce paperwork requirements, to improve permitting, and to promote waste reduction through recycling and reuse efforts, rather than to prevent pollution.

Three years after EPA Administrator Carol Browner announced the initiative, CSI has resulted in roughly 40 pilot projects, none of which have resulted in regulatory reform. Of these, eight are in the EPA's "electronics and computer sector group." Among the first was a project designed to

ensure that interested citizens could have access to EPA regulatory interpretations or determinations that are likely to change environmental management practices of target electronics and computer firms.

Frustrated with CSI's slow progress, the petroleum sector initiative col-
lapsed in 1996 after big companies defected. Automobile manufacturers
announced plans to withdraw from CSI by mid 1997 (*Inside EPA* 1997,
p. 1); however, the sector group completed its work and became inactive
midway through the year. The exodus of participants followed the departure of the representatives of several environmental-justice groups and all
the representatives from the state of Michigan.

Some members of the EPA's computer and electronics sector group have
attempted to address the limitations of CSI imposed by the current statutory framework by devising a sweeping alternative to the present system
of environmental laws and regulations. The result is a CSI project that
targets individual manufacturing facilities rather than industrial sectors.

The "Facility-Based Alternative System of Environmental Protection"
is a vision statement to provide companies regulatory flexibility for facilities "that commit to superior environmental performance" (US EPA
1995c). The statement calls for the development of an environmental
management system approach that sets performance objectives. Facilities
that participate in an alternative system would commit to continuous improvement in reducing both community and worker exposure to harmful
chemicals. The vision statement also seeks to integrate environmental,
health, and safety (EHS) programs into product design and production
processes.

Perhaps most central to the vision statement is the concept of "proportionality." According to the sector team's definition, proportionality
means that facilities seeking substantially increased operational flexibility
demonstrate equal EHS improvements. The improvements must exceed
regulatory standards and thresholds of the traditional system of environmental laws. Unveiled during CSI's first year, the alternative system project
was largely eclipsed by the Clinton administration's Project XL, unveiled
in 1995.

Related to CSI's legal limitations is the lack of adequate administrative
mechanisms and operating procedures. Primarily, CSI must compete for
staffing and resources with the EPA's air, water, waste, and toxics pro-

grams, each of which administers legal mandates and many of which face judicially imposed deadlines. Because Congress did not appropriate resources to the EPA to specifically administer CSI, the EPA must redirect resources from legally mandated programs to run CSI and similar initiatives that are not required by law.

A major procedural issue is the lack of methods to guide decision making. CSI architects envisioned that the initiative would generate quick "win-win" outcomes, with all participants walking away satisfied. It was hoped that participants would work out their differences to achieve consensus or unanimous approval of sectoral projects to improve regulation. However, the task of persuading parties that act as adversaries under the current system of laws to come to consensus proved problematic, and the EPA is now considering moving away from a strict interpretation of the term "consensus." The EPA may seek to relax the term to suggest less than 100 percent agreement.

Some of the earliest problems stemming from the lack of operating procedures were manifest in the computer and electronics sector. In 1996, lingering ambiguity over participation and decision-making authority resulted in the EPA's dismissal of two representatives from nonprofit organizations—one in the electronics and computer sector and two in printing—who were perceived by the agency as unduly obstructionist to the CSI process. At the same time, several representatives of environmental-justice groups resigned, maintaining that their organizations were underrepresented and in some cases unheard in CSI (US EPA 1996h).

Conversely, companies that made the good-will gesture of participating in CSI were reluctant to divulge potentially sensitive production and business information, in part because they feared that regulators would use the data to extract further concessions. Companies were furthermore concerned that nonprofit organizations might use information divulged during CSI meetings to mount lawsuits (NAPA 1994).

Industry's reluctance to disclose information about products and processes fueled mistrust among environmental, labor, and environmental-justice groups. During CSI's formative stages, the chief complaint of public interest groups in the computer and electronics sector was that they lacked the technical ability and the resources to effectively evaluate and critique industry proposals (ibid.).

Related to the problem of inadequate technical information is lingering ambiguity as to what actually constitutes the electronics and computer sector. CSI architects never concretely defined what they meant by the term "sector." It remains unclear whether CSI's computers and electronics sector refers broadly to all firms that fall under the two-digit Standard Industrial Code or, more narrowly, to manufacturers that make computers and electrical parts such as semiconductors.

In practice, CSI membership in the computer and electronics group is skewed toward a few very large firms, such as Texas Instruments, IBM, Intel, and Lucent Technologies. Public interest group participants in the electronics and computer group include representatives from the Silicon Valley Toxics Coalition and the Santa Clara Center for Occupational Safety and Health. To the extent that Intel, Lucent, and IBM have large operations and world-class EHS departments, these relatively integrated firms fail to represent the changing configuration of production networks among microchip manufacturers. Though it is tempting to conclude that large, integrated firms such as Intel represent the industry norm, the twin forces of economic restructuring and globalization make such firms look more and more like outliers. As was discussed in chapter 1, in terms of employees the chip firms in the United States are split evenly between large firm and small firms. Small firms include fabless companies that enlist foundry partners to produce chips. Furthermore, while companies such as Intel continue to build and operate wafer fabs independently, firms increasingly build fabs and manufacture products together. Also missing from the table were representatives from the software industry, which is integrated into the chip and hardware industries in that it fuels the demand for new chips and more powerful personal computers. Finally, missing from the CSI discussions was a discussion of how firms in the United States are increasingly building fabs and performing manufacturing and assembly functions in other parts of the world.

Owing to the Common Sense Initiative's poorly defined parameters, participants are largely unable to raise as issues tracking, accountability, and the geographic shifts that have resulted from restructuring and globalization. CSI focuses on issues associated with a few big firms, largely within the continental United States. It largely ignores the thornier problem of how current laws and policies fail to account for environmental problems associated with the operations of US companies abroad.

Despite such drawbacks, the computer and electronics sector has made some inroads. For example, a group attempting to assess what types of information firms should be required to report is making good use of an ongoing project by which the Texas Natural Resources Conservation Commission (TNRCC) is streamlining and consolidating all its environmental reporting requirements for electronics and computer firms. The original TNRCC project has been broadened to address the related question of what types of information interested citizens most need to assess environmental progress. Based on this assessment, the pilot group will design a "Combined, Uniform Report for the Environment," which will consolidate information that is currently collected under separate air, water, waste, and toxics laws. The computer and electronics sector group also is working with regulators and interest groups in Arizona to develop a comprehensive emergency response system that would consolidate separate emergency response planning requirements under the Emergency Planning and Community Right-to-Know Act and would help to better avert dangers posed by toxic materials common to electronics and computer firms.

Although CSI's legal and administrative barriers are considerable, they fail to fully account for why the effort has not advanced the initiative's non-regulatory goals, such as promoting pollution prevention and new environmental technology. Instead of focusing on pollution problems that arise at the front end of production, most of the sector projects currently underway focus primarily on product recycling, reporting, and developing alternative strategies for regulation. For example, the electronics and computer sector is examining regulatory barriers that discourage recycling of the cathode ray tubes used in television screens and computer monitors. Federal law typically defines used CRTs as hazardous, making it difficult to recover and recycle materials from them (such as the lead that shields users from harmful radiation).[2]

The computer and electronics sector also has launched a project to identify existing regulatory barriers that could prevent companies from eventually achieving zero wastewater discharge. Owing to the mounting

2. When handled or disposed of improperly, lead can impair mental function—particularly in children. Firms that manufacture or recycle electronics and computer products increasingly maintain that they are able to safely remove and recycle lead from CRTs.

challenges associated with water supply and wastewater disposal, a number of computer and electronics facilities are seeking methods of treating wastewater at the site where it is generated in order to recycle it back into production.

To date, the electronics and computer sector's most prominent achievement has been to focus greater attention on the growing problem of what one writer has described as the "inverse" of Moore's law: in theory, hardware sold 6 years ago has 1/16 the processing power of machines sold today (Parks 1997). As chip companies release newer, more powerful microprocessors and memory chips, computers and other electronic devices become obsolete more rapidly. Researchers at Carnegie Mellon University reported in 1991 that more than 2 million tons of computers will be sent to US landfills by the year 2000 (Navin-Chandra 1991). Researchers expect the disposal of computers to increase as semiconductor manufactures continue to release faster chips. The Carnegie Mellon study predicts that by 2005 a computer will become obsolete just as fast as a new computer is made (NSC 1997, p. 9). The Carnegie Mellon researchers furthermore estimate that up to three-fourths of the computers manufactured to date have not been disposed of but instead have been sent to storage facilities.

Most personal computers are retired in 3–4 years, according to Forrester Research (Parks 1997). Just as the "blue book" value of an automobile is said to drop the minute a buyer drives it off the lot, a new personal computer loses half its value within the first 12 months of ownership. After 4 years, a computer is worth only 5 percent of its original purchase price.

According to the Gartner Group, a marketing and research firm in Stamford, Connecticut, about 79 million computers had been retired from their primary lives in 1996. By 1999, another 42 million machines will be retired. "Retired," researchers have found, is a polite euphemism. To date the Gartner Group has found that about 65 percent of the retired machines owned by businesses end up in storage—squeezed into office closets with the coffee creamer and cleaning suppliers, under desks, or in warehouses. About 15 percent are resold, and another 15 percent are dismantled for scrap or recycling. About 5 percent find their way to schools and to nonprofit organizations that either rebuild and redistribute the machines or use them in house.

In 1997, the CSI electronics and computer sector sponsored an Electronic Product Recovery and Recycling Conference. More than 230 people attended the event, which focused on topics such as developing an infrastructure of suppliers to recycle equipment, allocating responsibility for used equipment, and promoting strategies to design products that can be easily disassembled and recycled. Participants also discussed how regulations such as definitions of what constitutes a hazardous waste can discourage the recycling of cathode ray tubes (NSC 1997).

American companies with operations in Germany and the Netherlands are already required to "take back" products once their useful life has ended (Dillon 1994). In anticipation that some may lobby Congress to pass similar laws, the conferees explored the use of alternative instruments to promote product reuse, such as additional tax incentives for charitable donations of used computers and electronics equipment.

The CSI electronics and computer sector used the conference to unveil preliminary findings from a study that explored the feasibility of community electronics and computer recycling. The study, which targeted three counties in New Jersey, was designed to determine the composition of the waste stream and the economic viability of collecting, sorting, and selling electronic refuse to firms that specialize in recycling and reuse. The researchers found that televisions, audio equipment, toasters, and air conditioners, rather than used computer equipment, constituted most of the waste. Consistent with the high proportion of electronic products in the waste stream, the researchers found much higher rates of participation among private households than among businesses (Stanaback 1997, pp. 3–8). They also found that the prospects for promoting the expansion of salvaged material markets are dim, because secondary material prices (and hence potential profits) are very low. For example, a typical 386 personal computer has a resale value of about $10–$20 and a component value of $5. Four megabytes of used memory chips fetch about $1. The value of the metal inside is about 50 cents (NSC 1997, p. 25). Clean used plastic currently fetches between 5 and 30 cents for roughly 500 pounds. Thus, the net value of the plastic and the glass in an individual machine is virtually zero. The accelerating rate of new PC introductions will cause prices for PC scrap materials to drop even further.

It is possible that low scrap material prices may benefit the small but growing segment of "chip pickers"—refurbishers who rebuild and resell used computers. According to the *Wall Street Journal,* 2.4 million used computers were sold in 1996. At present, however, few first-time buyers are willing to purchase used machines. Even though many of the machines sold by refurbishers come with warrantees, first-time buyers and the "technically challenged" may find it far easier to plunk down the money for a new machine than to locate parts for yesterday's drives and motherboards and to find versions of software that are no longer stocked on store shelves. As the columnist Liza Mundy quipped in the 13 July 1997 *Washington Post:* "Clearly this sort of thing is no accident; new technologies are not introduced because they are better, but also because manufacturers want to induce in us the urgent need to renew, to upgrade, to buy and buy again."

Failure to Tackle Dynamism

The themes of the aforementioned conference, consistent with most pilot projects advanced by the CSI electronics and computer sector group, underscore the challenge of successfully addressing rapid chip innovation. Though most agree that the rapid release of computers and other electronic products poses new environmental problems, few seek to examine how waste problems are shaped by consumption patterns, which ultimately are influenced by the semiconductor industry's rate of new-product introduction. The experience of driving a 5-year-old automobile is, for the most part, the same as that of driving a new one. However, owing to chip advances, a 5-year-old personal computer simply fails to navigate the information superhighway in the same way as the latest IBM or Apple machine. Although from the outside today's PC does not look much different from one made in 1992, in terms of the computing functions made possible by its microchips it is an entirely different product.

Clearly, product design is only one aspect of the computer waste problem. An even bigger aspect of the problem stems from consumers' preferences—something that federal laws typically refrain from targeting. As the experience of CSI shows, the consumption question is so sensitive that to raise it in the context of a voluntary initiative is virtually impossible. Bette Fishbein, a panelist at the 1997 Electronic Product Recovery and Recycling Conference, boldly challenged the federal government to adopt public policies to "shift the economic trajectory" of "companies [that] make most of their money by rapid product change and rapid product obsolescence." Not surprisingly, her bid was met mostly with silence from the audience, most of whom preferred instead to pursue end-of-pipe control strategies such as recycling and reuse. (It might have been more illuminating to poll the participants on the age of their own computers.) Not surprisingly, the conference's "breakout sessions" primarily sought to identify strategies to spur the development of secondary markets for used computer and electronic products and scrap material rather than raising the much more vexing question of whether individuals and companies around the world are willing to stop purchasing the latest, fastest PCs (NSC 1997, pp. 20–25).

Although recycling and reuse may be the most palatable and least economically disruptive policy alternatives, how the industry's dynamism may undermine the viability of such strategies has not been addressed sufficiently. As sustained chip breakthroughs continue to accelerate the obsolescence of computers and electronic products, the mounting waste pile promises to further depress prices for used equipment and material (ibid., p. 25).

The conference participants' reluctance to tackle consumption questions was consistent with CSI's poorly defined "sectoral" focus. The initiative has largely failed to establish the degree to which chip, computer, and electronics firms are intertwined. New chips lead to new machines and software applications. By failing to concretely define what constitutes the electronics and computer sector, CSI missed a tremendous opportunity to advance the debate on appropriate environmental management strategies for this set of industries. The lack of a refined framework in which to analyze the structure and locational features of chip, computer, and electronics firms has resulted in a set of pilot projects that fail to venture beyond the current control-based focus of environmental laws and beyond the territorial boundaries of the United States.

Summary

By moving the focus away from individual laws and toward industry sectors, the Common Sense Initiative was designed to enable participants to develop regulations that would result in "cleaner, cheaper, smarter" environmental management strategies. To date, the initiative has resulted in no regulatory changes and few prevention efforts. In particular, the initiative has suffered from the EPA's naive expectation that years of adversarial relationships among business, environmental, and labor groups could be solved overnight. J. Clarence Davies, a longtime observer of federal environmental initiatives, put it this way: "Programs which depend for their success on cooperation, voluntariness, and trust still do not fare well." (Davies and Mazurek 1996, p. ii) Because CSI is a "designer" initiative rather than a program backed by law and by budget dollars, the EPA from the outset lacked sufficient resources to convene, facilitate, and provide technical support to CSI's numerous constituents. From the industry's perspective, CSI provides few incentives for firms to disclose information about pollution problems at their facilities. Furthermore, the slow and often contentious nature of the initiative has led firms to question whether the costs of participation are worth the benefits.

To provide a few companies with a more rapid mechanism for flexibility, the EPA developed Project XL. In theory, Project XL and CSI have the potential to be complementary. Instead, Project XL has been perceived by some CSI participants as a competitor. As will be discussed in detail in chapter 6, Intel's 1995 decision to simultaneously participate in both Project XL and CSI ultimately hampered trust-building among CSI's participants (which included Intel and the Silicon Valley Toxics Coalition).

Despite its administrative shortcomings, the Common Sense Initiative has offered the EPA, industry, and interest groups a tremendous opportunity to construct a sectoral profile that would promote prevention and would more fully illustrate the environmental challenges associated with the industry's dynamism. Since participation has been skewed toward a few prominent companies, CSI has furthermore failed to address the environmental challenges associated with modifications in where and how firms manufacture computers and the chips that power them.

The EPA has missed in CSI an opportunity to develop data that could supplement the parallel yet separate Project XL, which focuses not on sectors but exclusively on facilities (including an Intel fab). A central source of controversy in regard to Project XL was the lack of sufficient information to assess the merits of Intel's proposal and to assure participants in XL that the firm would meet environmental and economic goals. CSI could have been an opportunity for identifying in advance what types of technical information are indeed necessary for Project XL's public participants to weigh the tradeoffs of facility-based regulatory experiments.

CSI's sectoral focus also provided the EPA with an opportunity to compare the environmental and economic progress of XL participants with that of firms that continue to follow traditional command-and-control approaches. Furthermore, CSI could have provided a way to popularize and transfer facility-based experiments such as XL to other companies.

Finally, the members of the CSI computer and electronics sector, who had a one-year lead on Intel's XL participants in studying issues specific to a handful of prominent computer and electronics firms, could have been enlisted at Intel's XL site to help minimize charges that national interest groups were locked out of Intel's Project XL proceedings.

6

The XL Files

To business this administration is saying: If you can find a cheaper, more efficient way than government regulatory requirements to meet tough pollution standards, do it—as long as you can do it right.
—President Bill Clinton, State of the Union Address, January 1996

In November of 1996, EPA Administrator Carol Browner joined Intel officials at the company's new production site near Phoenix to launch Project XL—the "crown jewel" in the Clinton administration's effort to reduce regulatory burdens to business. The 5-year project agreement covers operations at Intel's 720-acre Ocotillo site in Chandler, Arizona.

Since Intel first rode into Chandler, in 1980, serpentine ocotillo plants and orange groves have increasingly given way to high-tech facilities, high-paying jobs, and new housing developments. Locals credit Intel with transforming this former agricultural stop into a high-tech boomtown. According to an article in the 22 October 1995 *Arizona Republic,* Chandler is the nation's third-fastest-growing city of more than 100,000.

The Ocotillo agreement, negotiated among Intel, regulators, and five local residents, gives the world's largest chip maker the ability to make routine process changes at Fab 12, a new Pentium microprocessor manufacturing facility. Furthermore, the XL air permit—good for 5 years—allows Intel to add a second wafer fabrication facility at the Ocotillo site without securing additional air permits. The permitting process can take from 6 to 12 months. As of June 1998 Intel had not announced plans to build a second fab at the Ocotillo site, but it had added 30,000 square feet of clean-room space to Fab 12. In exchange for the ability to modify or expand manufacturing without securing additional permits, Intel pledges to keep air emissions from the site below what current federal laws require.

Although Intel and official Project XL stakeholders maintain that the agreement offers protection that is equal if not superior to the protection required under current laws, environmental groups call the Project XL agreement a "sweetheart deal" for industry. The Campaign for Responsible Technology (an international counterpart of the Silicon Valley Toxics Coalition that links environmental, labor, and environmental-justice groups that work on high-tech issues) collected more than 130 signatures on a petition that opposed Intel's XL plan. Public interest groups have criticized Project XL's technical basis and its policy basis (CRT 1996a). Opponents of the Project XL effort argue that the agreement fails to provide sufficient assurance that Intel will (to paraphrase President Clinton) do it right.

National environmental groups and environmental-justice organizations question the EPA's decision to exclude them from the XL negotiations, which involved only local residents. The non-local organizations, several of which participated in the EPA's parallel Common Sense Initiative, charge that the XL agreement fails to deliver environmental performance superior, and in some cases even equal, to what current federal laws require. Groups from outside Chandler also criticize the EPA for failing to use its leverage to push the world's largest microprocessor producer to economize on its use of water and chemicals.

This chapter describes the genesis of Project XL to illustrate the challenge of designing policies to better address rapid change in semiconductor manufacturing. Like CSI, Project XL fails to define adequately who may participate and how environmental progress will be measured. The EPA also underestimated the time it would take for groups that act as adversaries under the present system to overcome their mutual mistrust, throwing into question whether an XL-type process really saves time and money for a fast-moving company such as Intel.

In addition to procedural problems, Project XL also has been plagued by the lack of legal mechanisms to protect potential participants. As envisioned, Project XL would provide firms with the ability to violate current regulatory standards in exchange for superior environmental performance. For companies, the risk of enforcement action and citizens' suits associated with XL appeared to be too high. Furthermore, since the majority of the EPA's activity is driven by laws and court-imposed deadlines, the lack of legal authorization for XL meant that the agency did not have sufficient resources to administer the initiative. On 30 October 1997, Senator Joseph Lieberman introduced legislation that would authorize XL-type projects (Senate Bill 1348). The Innovative Environmental Strategies Act of 1997 would reduce the legal risks to companies of participating in XL-type initiatives and may also provide the EPA with adequate resources to administer such programs. Environmental groups maintain that the proposed legislation does not adequately address lingering uncertainties about who may legitimately participate in XL-type efforts and what monitoring methods should be used to ensure that environmental goals are being met.

Project XL's considerable administrative and legal weaknesses were severely compounded by the EPA's decision to use as a "pilot" an industry in which production processes and potential environmental problems change rapidly and thus remain poorly understood by public officials and public interest groups. Although Project XL is designed to promote superior environmental and economic performance, a number of environmental groups maintain that not enough information exists to determine whether the Intel plan is in fact warranted, as well as what the ultimate economic and environmental effects of the experiment might be. As envisioned, XL projects, if successful, would be transferred to similar firms and perhaps even entire industries. Such goals are unlikely because the XL effort first failed to account for the highly heterogeneous nature of chip fabs, firms, and markets—not to mention the enormous leverage that a multinational company may have over an individual town and its residents.

Origins of Project XL

Whereas the Common Sense Initiative was based on the idea that regulations should recognize differences among groups of manufacturers (termed "sectors" by the EPA), Project XL was premised on the concept that, even within a sector, equipment, process chemicals, and the capacity to control pollution vary among facilities that make the same types of

Table 6.1
Goals of Project XL. Source: US EPA 1996c.

Environmental results
Cost savings and paperwork reduction
Stakeholder support
Innovation/multi-media pollution prevention
Transferability
Feasibility
Monitoring, reporting, evaluation
No shifting of risk burden

products. For example, Amoco was required to control volatile organic compounds from its waste treatment plant (a minor source of emissions) rather than to control emissions from a refinery's barge unloading facilities (the second-largest source of volatile organic compounds at the plant). Industry, regulators, and public representatives who studied the plant discovered that controls on the barge equipment, coupled with a few other changes, could have curbed 97 percent of Amoco's VOC reduction requirements at one-fourth the cost of the waste treatment plant controls (Amoco and US EPA 1992, p. 11).

In order to stimulate more projects that lower costs and cut pollution, the Clinton administration announced the XL program in a 1995 policy statement (Clinton and Gore 1995). Shortly thereafter, the EPA developed enforcement policies and protocols to solicit XL proposals from industry and government jurisdictions. At the time, the EPA's primary criterion for acceptance into the XL program was that firms demonstrate environmental results "better" than they could achieve under full compliance with the present laws and regulations (US EPA 1996c). "Better" was loosely defined on the assumption that XL was a work in progress and that different definitions of what constitutes such performance would emerge as the program matured. The second major criterion was cost savings. Such results are to be achieved by largely proposals that take a more holistic, prevention-oriented approach to reducing pollution. Because XL relies on added public oversight, the EPA favored proposals that received strong endorsement from the communities in which plants were located. The EPA also tried to solicit projects that had a high potential to be emulated elsewhere (ibid.). (See table 6.1.)

The EPA gave firms accepted into the program 6 months to develop and negotiate a final project agreement that sets out the terms and conditions of how a facility operates under the initiative. As part of the development process, the EPA required XL participants to establish a "baseline" level of pollution before implementation and compare environmental performance after implementation. The EPA mandated that part of the "baseline" calculation include the level of performance that the facility would attain under "full compliance" with the current statutory system. The burden would then fall upon the participating facility to demonstrate to the EPA and community residents that the underlying assumptions of the calculation are correct.

Since 1995, the EPA has approved six final project agreements that are now underway. The first involves a permit that consolidates all federal, state, and local environmental requirements into one document for a Florida citrus juice processor. Consolidating permits eliminates the requirement of preparing multiple applications, a benefit that could save the company several million dollars. Though it was approved after the juice plant's consolidated permit, Intel's effort represents the first such agreement awarded to a major company with manufacturing operations throughout the United States and, increasingly, around the world. The remaining participants include four facilities in New York and New Hampshire operated by HADCO, a printed wiring board manufacturer; a Weyerhaeuser pulp and paper manufacturing plant on the Flint River in Georgia; a Merck pharmaceutical plant in Virginia; Osi Specialties, a West Virginia chemical manufacturer, and Vandenberg Air Force Base in California. An additional 15 projects are in various stages of development, and 30 proposals have been withdrawn by the applicant or rejected by the EPA (US EPA 1998).

Legal and Administrative Drawbacks

To help distinguish XL from other agency projects, some EPA employees coined the expression "If it isn't illegal, it isn't XL." That motto, circulated among XL project staff in an agency update dated 11 March 1996, was to help distinguish Project XL from other regulatory experiments, such as the Common Sense Initiative. As envisioned, XL would allow firms to seek waivers from laws and regulations to achieve cleaner and cheaper environmental results. Unfortunately, XL's strength was its primary weakness: In the absence of statutory permission to break the law, firms out of compliance are vulnerable to EPA enforcement action and to citizen suits.

In theory, greater public participation in projects such as XL would diminish potential legal challenges, because all effected parties would be required to negotiate and approve agreements. However, at the outset XL failed to contain specific procedures that set out exactly who could participate in project outcomes and to what degree public participants would influence those outcomes.

Before it became apparent to both the EPA and industry that the risks to companies of seeking regulatory flexibility would be too high, industry submitted to the EPA some daring proposals that could control more pollution at lower cost. For example, one company proposed to use a new technology that it claimed could recycle hazardous waste into glass. The applicant sought regulatory relief from hazardous waste law requirements in order to test its new recycling process. The EPA rejected the application because the project appeared too risky (Lavelle 1996). Another applicant sought to relax the ban on imports of polychlorinated biphenyls (PCBs) so that it could accept materials containing the substance into the United States from Canada and Mexico. Intel initially proposed a multimedia operating contract that would have allowed the facility to emit more pollutants that are expensive to control in exchange for controlling more pollutants that are less expensive to prevent or control. In other words, if air pollution abatement methods are more costly than water pollution controls, the company could emit more air pollution if increased air emissions were offset by additional water pollution controls (US EPA 1995a).

The EPA failed to approve the hazardous waste and PCB proposals but approved a modified version of the Intel plan. Intel officials abandoned the idea of seeking regulatory flexibility to conduct cross-media emissions trades. Instead, Intel managers decided to develop a project that would comply with all current federal, state, and local legal requirements. Though such a policy would minimize the risk to Intel of a legal challenge, the lack of flexibility diminished the potential economic benefits of participating in Project XL.

Intel's Motivations

Project XL's premise has the enormous potential to improve how regulations target a few leading-edge wafer fabs, where production chemicals and equipment vary not only among facilities but also over time at an individual facility (Hatcher 1994; Sheppard 1995). Intel routinely doubles the number of transistors on a piece of silicon every 18 months. To achieve constant refinements, Intel may modify its process chemistries 30–45 times per year. For example, if tests show that the use of one substance (say, hydrogen) excessively damages a wafer and causes malfunction, Intel may seek to substitute another chemical, such as chlorine. In addition to chemical changes, Intel modifies both the type and location of equipment on average between one and three times per year (Hatcher 1994, pp. 7–8). Though changes are necessary to reduce defective products, modifications may vary both the type and quantity of conventional and hazardous air pollutants (HAPs) generated through wafer fabrication.

To better control air pollution, the 1990 Clean Air Act Amendments require each state to develop a comprehensive, federally enforceable operating permit program for all major stationary sources of air emissions. "Major sources" may be entire facilities or individual pieces of equipment that emit more than a specified number of tons per year of the six conventional pollutants and those defined as hazardous air pollutants. States may require that "minor sources" be subject to modification review.

Intel's success is a partial function of the firm's commitment to continuous product innovation. More than 80 percent of all personal computers in the world are powered by Intel microprocessors. Sales in 1996 were $20.8 billion, making Intel one of the top five earners of the Fortune 500 (Kirkpatrick 1997, p. 62). Intel contended that the new federal air permitting requirements had the potential to harm its competitiveness. The provisions require companies to obtain 5-year air permits that establish emissions limits. Since states have considerable latitude in devising permitting programs, some require companies to merely notify regulators when process changes occur. Other states may require that companies obtain approval for each process change, which could impose delays of several days or months, depending on whether regulators seek public comment.

Of concern to Intel is the potential for air permits to stall manufacturing and the release of new products. As was discussed in chapter 1, the chip maker who is able to produce in high-volume chips ahead of competitors dominates the markets and earns higher-than-normal profits (Gruber 1994). Competitors who eventually release and sell the same product at a lower price are unable to earn the same profits. Air permits may not pose much of a problem to chip makers whose profitability is not tied to constant innovation, or to industries where processes and equipment are relatively stable. However, the Clean Air Act Amendments' permitting provisions caused Intel to "seriously question whether it could remain committed to the construction and expansion of our US sites" (Hatcher 1994, p. 4).

To ensure that sufficient production capacity exists for new products, Intel often designs billion-dollar fabs and secures permits for them several years in advance of full operation. In order to secure permits for wafer fabs still years away from shipping chips, Intel engineers develop estimates based on manufacturer's specifications. Equipment is then initially tested on comparable—but not identical—processes at other fabs. Because production processes and equipment are constantly modified, the potential for miscalculation is relatively high.

In the case of Fab 12 in Chandler, Intel applied to the Maricopa County Environmental Services Department for an operating permit in 1993, 3 years before the facility would actually begin shipping advanced Pentium microprocessors. Because the fab was not yet in operation at the time of the initial permit application, Intel engineers developed estimates that showed the facility would emit 5.5 tons of hazardous air pollutants per year. That level was sufficiently high to cover any emissions that would result from experimental phases of production before the fab started manufacturing chips for market.

As Fab 12 approached full output in 1996, engineers found that the scrubbers only removed 20–30 percent of pollutants. As a result, Intel sought in its XL permit to release 20 tons of HAPs—4 times what was allowed in Intel's original air permit, issued in 1994. According to Intel, the lower scrubber efficiency rates were due to the fact that the manufacturing processes generated fewer volatile organic compounds than originally forecast. Engineers discovered that the scrubbers were less effective at removing low concentrations of volatile organic compounds than at removing high concentrations. Another reason that the estimates were off is that engineers subsequently added large amounts of methanol (a HAP) to the manufacturing process.

In practice, the potential for Intel to experience delays due to air permits at fabs where chips are already in production is minimal for two reasons. First, although Intel feared that major source permitting provisions of the 1990 Clean Air Act Amendments would harm competitiveness, it nonetheless adopted a corporate policy of structuring almost all of its US facilities to emit pollutants well under the threshold limits that would trigger reporting for major sources. One exception is a facility in Aloha, Oregon, that houses some older production lines and is a major source of VOCs. There, Intel worked with the state of Oregon and the US EPA to develop a "Pollution Prevention in Permitting Pilot" (P4) permit that placed a plant-wide cap on emissions and fully enforceable pollution prevention requirements. In exchange, Intel received an allowance that approved in advance routine production changes at the older fab. The P4 permit conditions require Intel to describe what types of processes and decision making the facility will use to reduce pollution. The P4 permit also requires Intel to establish partnership agreements with its suppliers and equipment vendors to select raw materials that minimize hazardous air pollutants and volatile organic compound emissions. Finally, the permit requires Intel to develop a data collection system and an employee training program to promote pollution prevention (ODEQ 1994).

The risk of delays to chips in production is further minimized in regions that support high concentrations of chip firms. In both California and Arizona, regulatory agencies have developed permitting provisions that recognize the industry's unique production characteristics. For example, the Bay Area Air Quality Management District and Maricopa County have not applied minor new source notification requirements to the constant process adjustments and modifications necessary to refine products (Hatcher 1994, p. 9). Thus, the delays ostensibly of most concern to Intel involve the plant expansions to accommodate the manufacturing of new, rather than existing microchips. Both the Aloha and Ocotillo air permits allowed Intel to not only make routine process changes without seeking permit re-approval but expand and—in the case of the Ocotillo site— build an additional fab.

The Intel Project XL Proposal

Intel's XL agreement at the Ocotillo site is much broader than the Oregon effort, which simply focused on air permits. As part of its XL effort, Intel proposed to develop an "environmental master plan" to cover operations not only at Fab 12 but the entire 720-acre Ocotillo site, including the second possible fab referred to by Intel as "Fab X."

As required by the EPA, the terms of the plan were developed jointly by representatives from Intel, ten different agencies, and five Chandler residents. Though the process was originally slated by the EPA to require 6 months, Intel's master plan took over 11 months, 100 official meetings, and dozens of other informal conversations and meetings to develop (Coombs 1996). Meetings, held mostly inside the plant, were not open to the public, but people interested in attending could request to attend by submitting a written request in advance. In order to attend, all visitors were required to clear security and sign a confidentiality pledge not to disclose any information learned at the site. Both Intel and the EPA posted minutes from a dozen public meetings on their respective web sites (US EPA 1996e; Intel 1996b).

As originally envisioned, the XL agreement would have allowed Intel to consolidate permitting and reporting into a single document administered by the state of Arizona rather than by ten separate agencies. The master plan also contained features of interest to Chandler residents, most of which were not addressed by federal laws. Chandler residents were mostly concerned with the plant's impact on water supplies and air quality (Coombs 1996). For example, the plan contained a provision to recharge the plant's wastewater into the groundwater through the construction of a $28 million purification plant. That plant, constructed before the XL negotiation, uses reverse osmosis to treat effluent to a point where it is at least as clean as drinking water. At those levels the water is too dirty for microchip manufacturing. However, the new purification plant makes it possible for Intel's water to be re-injected into the local aquifer or used in irrigation. Intel's plan also contained commitments to reduce solid waste and employees' vehicle use.

Intel expected the master plan to save on reporting costs. As part of its provisions, the plan would have consolidated all environmental reporting and permitting requirements and placed responsibility for their oversight with the Arizona Department of Environmental Quality (Intel 1996b).

The Final Project Agreement

The final project agreement, signed in November of 1996 by Carol Browner, contains most of the elements of Intel's master plan proposal. However, Intel was unable to vest permitting and reporting requirements with a single agency, because the Maricopa County Environmental Services Department, the local agency responsible for air pollution control, refused to relinquish its air permitting and oversight authority to the state of Arizona.

The final project agreement includes a permit from the city of Chandler that allows Intel to recharge the local aquifer with treated wastewater; it also includes a number of voluntary and mandated requirements of interest to the Chandler community, such as water conservation and air quality issues beyond the plant's fence line (US EPA 1996f). The agreement also contains pledges outside the immediate sphere of the environment. For example, it requires Intel to donate computers and equipment to local schools and libraries.

The air portions of the agreement are contained under a separate, enforceable 5-year permit (US EPA 1996a). Under the XL permit, Intel pledges to cap emissions of conventional and hazardous air pollutants for Fab 12 and a possible second fab below what federal laws require for a facility to remain a minor source. Table 6.2 contrasts emissions limits for Fab 12 and an additional facility with both federal emissions limits for minor sources and emissions levels allowed under the 1994 permit for Fab 12 issued by Maricopa County.

Emissions of conventional pollutants under the XL permit are well within the limits allowed by federal law for minor sources. The limits for HAPs under XL are slightly less than current federal limits for minor

Table 6.2
Minor source limits, 1994 permit levels, and XL permit levels. Source: US EPA 1996a.

Pollutant (tons/year)	Federal requirements for minor sources	1994 Fab 12 permit	Project XL permit, Fab 12 + Fab X[a]
Carbon monoxide	<100	59	49
Nitrogen oxide	<100	53	49
Sulfur dioxide	<250	10	5
Particulates 10 μm or smaller	<70	7.8	5
Total volatile organic compounds	<100	25	40
Hazardous air pollutants[b]	<25 aggregate; 10 for any individual HAP	5.5	10 total organic 10 total inorganic

a. Though Intel has not officially announced plans to construct a second fab at the Ocotillo site, the emissions levels under the XL permit column are for two fabs.
b. Hazardous air pollutants (HAPs) are those listed in section 112(b) of the federal Clean Air Act, as amended. The 10 ton per year limits for total organic HAPs and total inorganic HAPs assume that more than one HAP will be emitted from the site. If a single HAP is emitted from the site, the emissions limit is 9.9 tons per year. On the basis of Intel's modeling exercise and the Arizona Ambient Air Quality Guidelines (AAAQG), the permit establishes separate limits for phosphene (4 tons per year) and sulfuric acid (9 tons per year), to be included in the aggregated combined inorganic HAP emissions plant site emissions limit.

sources. The XL air permit specifies that Fab 12 and any other additional operations combined may not emit more than 20 tons a year of HAPs. Of those 20 tons, no more than 10 tons may be organic and 10 tons inorganic.

Though the XL emissions levels compare favorably to federal requirements, they compare less favorably to Intel's original 1994 air permit. Specifically, HAP emissions under the XL permit appear to increase fourfold. The increase is the result of lower-than-expected removal efficiencies on pollution control equipment at low inlet concentrations.

To ensure that hazardous emissions levels under the XL permit were sufficient to protect human health, Intel conducted fate and transport modeling to ensure that emissions levels were below those specified by a set of voluntary risk-based guidelines on 400 chemicals developed by the state of Arizona. While the XL permit limits hazardous air emissions to 10 tons per year for all organic pollutants and 10 tons per year for all inorganic pollutants, Intel based the modeling exercises on the conservative assumption that each individual pollutant would be emitted at a rate of 10 tons per year.

Intel's analysis showed that all but two pollutants (phosphine and sulfuric acid) would meet or exceed Arizona's risk-based guidelines. Based on the results, the XL permit establishes separate limits for phosphine (4 tons per year) and sulfuric acid (9 tons per year), to be included in the aggregated combined inorganic HAP plant site emissions limit. With the exception of phosphine and sulfuric acid, the XL permit's aggregate cap provisions for organic and inorganic HAPs make it unnecessary for Intel to specify individual substances emitted as long as Fab 12 and any additional facility remain under the limits specified in the XL agreement.

Although it forms just one part of the final plan, the air permitting portion of the XL agreement was the most prominent and contentious element. Several national advocacy organizations and several industry watchdog groups from electronics-producing regions outside Chandler faulted both the permit's technical features and its policy approach.

Cleaner Than What?

Of primary concern was the thorny technical question of how to measure environmental performance. The calculation of environmental benefits requires definition of a "baseline" level of emissions associated with conventional regulation. In other words, what would the facility release in the absence of Project XL? Benefits represent the difference between how much the facility could have released and how much XL actually reduced emissions. Unfortunately, since Fab 12 is a new facility and the second fab exists only on paper, the site lacks an emissions history with which to develop a base case with which to determine how much, if anything, the XL permit caused Intel to reduce emissions.

The lack of historic data left the "baseline" issue open for debate. Absent historic data, environmental groups said that Fab 12's performance under XL should be compared to emissions levels specified in Intel's original 1994 air permit. The EPA and other members of the Project XL stakeholder group maintained that the federal requirements for minor sources were more appropriate "baselines" (table 6.2).

One reason the "baseline" issue was so hotly contested is that Intel underestimated HAP emissions for its XL permit. Intel's error raised concerns among some community members in the XL air permit work group about the accuracy of the information provided by the company (US EPA 1996f, pp. 2–3).

Another reason the HAP portion of the XL air permit was controversial is that it allows Intel to aggregate pollutants rather than list them individually. Under XL, Intel can mix and match any combination of hazardous pollutants it likes, as long as total emissions remain under an aggregate plant-wide limit and remain within risk-based guidelines set by the state. Environmental groups maintained that not enough scientific evidence exists to measure what effect (if any) an aggregate cap might have on humans and the environment. Measuring the human-health and environmental effects of the HAP portion of Intel's XL air permit is currently beyond the capacity of even the most sophisticated risk models. In contrast to conventional pollutants (such as carbon monoxide or sulfur dioxide) that occur nationwide, HAPs tend to cause health problems that are more localized. The Clean Air Act Amendments address HAPs separately because they trigger effects at smaller doses than conventional pollutants. Furthermore, HAPs typically possess unique exposure thresholds, pathways, and properties once released into the environment: some are extremely toxic in small amounts whereas others require much larger doses; some act through exposure on contact, some on inhalation; others disperse rapidly in air whereas some persist in soil or water.

In the absence of an emissions profile for Fab 12, the Natural Resources Defense Council suggested that an independent group develop industry benchmark data that could be used to compare hazardous emissions from Fab 12 with emissions from similar facilities (NRDC 1996a). In response, EPA staff charged with administering Project XL attempted to locate a profile of HAP emissions for semiconductor manufacturers. EPA project staff discovered that the agency has not developed such data. According to the EPA (1996f), "the absence of benchmark data for the entire semiconductor industry makes assessments of superior environmental performance problematic."

To compare HAP emissions from Fab 12 with other facilities, the EPA analyzed total TRI emissions for six hazardous substances between 1992 and 1994 for similar-size facilities within SIC code 3674 (semiconductors and related devices). The EPA calculated an industry average of 18 tons of HAP emissions per year. Although EPA analysts added numerous caveats to underscore the basic limitations of TRI data, the EPA (1996f) used the information to conclude that Intel was "well within, if not exceeding, the standard for the industry." Such an analysis assumed, however, that emissions from a new, state-of-the-art microprocessor fab are comparable to emissions averaged over all facilities that use SIC 3674 as a primary reporting code. Such aggregates reflect emissions from all semiconductor facilities currently in operation—not just new, state-of-the-art fabs. Moreover, the EPA analysis reported that emissions from Ocotillo were merely "comparable" rather than superior to facilities that report to the TRI using SIC 3674 as a primary reporting code.

The XL stakeholders and the EPA countered that the emissions levels in the 1994 operating permit were artificially low because they covered only emissions resulting from experimentation, and not those from actual production. In contrast, the XL permit emissions levels apply to Fab 12 at full output. Moreover, the XL caps represent a combined limit that covers not only Fab 12 but also a possible second fab.

Despite the objections of national environmental groups, the EPA and the Intel XL stakeholders decided that, absent historic data or industry benchmark information, Intel's air emissions under Project XL need only fall within what would be required under reasonably anticipated future regulation. From this perspective, the minor source limits under the 1990 Clean Air Act Amendments become the appropriate performance gauge. According to this measure, then, the environmental benefits of XL represent the difference between emissions levels under the XL permit and the theoretical maximum allowed by federal law (table 6.2). For example, the XL permit reduces carbon monoxide and nitrogen oxide each by 51 tons per year.

To address risks under the HAP portion of the XL air permit, Intel's final project agreement, approved by the EPA, provides protection equal to that afforded by traditional permits. For example, to ensure that hazardous emissions levels under the XL permit sufficiently protect human health, Intel agreed to perform fate and transport modeling exercises to analyze the effects of HAPs that both fabs are likely to emit during the period 1996–2001. The XL permit requires Intel to conduct similar analyses in the future before any new regulated chemical listed under the Arizona guidelines or any unregulated compound is emitted from the facility.

However, lingering scientific uncertainties about the potential effects of hazardous air pollutants on humans and the environment prompted the Natural Resources Defense Council, the Silicon Valley Toxics Coalition, and several other organizations to question whether the plant-wide caps on hazardous emissions provide benefits that are superior to those under a traditional air permit. For example, the NRDC expressed concern that the permit contains no provisions to prevent Intel from using increasingly hazardous substances over time. Also, because the emissions limits are set on an annual basis, there is nothing to prevent Intel from emitting pollution "spikes" at intervals (NRDC 1996a,b; US EPA 1996f). The EPA countered that emissions levels under the XL permit for Fab 12, and those for any additional facility, were sufficient to protect human and environmental health because they were well within "minor source" designations under the Clean Air Act (US EPA 1996f). The EPA (ibid.) did not dispute the NRDC's observation that the XL permit contains no provisions to prevent emissions from the plant from becoming more toxic over time, but it maintains that Intel's conservative modeling exercises showed that even if the manufacturer were to release 10 tons of a single hazardous pollutant the levels would remain well Arizona's guidelines for ambient air quality.

Non-local environmental groups did successfully lobby the EPA to ensure that the increases in emissions from Fab 12 would not outstrip increases in production levels. The XL air permit allows Intel to increase its total loadings of air pollution over the 5-year permit life, as long as loadings do not exceed output levels (ibid.). To ensure that pollution loadings (as opposed to the composition of those loadings) do not exceed output, the permit requires Intel to work with the EPA to develop a measure of emissions per production unit. However, development of the production unit was not part of the XL negotiation process. Instead, the EPA and Intel agreed to develop the normalization measure after Carol Browner signed the XL agreement.

To help the public monitor emissions under the XL agreement, Intel agreed to release emissions reports from the facility each quarter. Emissions estimates are based on predetermined formulas developed by the EPA, based on the flow of materials and energy into and out of the fab. Such flows are estimated by emissions factors that consider fuel use and the type of equipment generating the pollution. However, Intel has claimed as confidential any emission factors that apply to specific tools.

Several non-local environmental groups unsuccessfully requested that Intel, as a party to the XL provisions, be required to do more than merely estimate emissions. The Silicon Valley Toxics Coalition wanted Intel to install emissions monitoring devices that would provide actual, real-time measures of air pollution. As was discussed in chapter 2, however, Intel and other chip manufacturers maintain that such devices are too costly and may introduce contamination.

To improve accountability, the permit stipulates that Maricopa County officials must inspect the facility and require the plant to verify the data in the firm's quarterly emissions reports. The county officials have authority to decide whether the data merit protection. In a briefing paper, Henry S. Cole, a meteorologist and a former EPA staff scientist, pointed out that the agreement contained no provisions to allow the public to monitor compliance or to verify what Intel reports as its emissions limits (Cole 1996).

Prevention or Control?

At Amoco's Yorktown refinery, participants in a study intended to develop more cost-effective methods of controlling pollution identified two control strategies for one pollutant and selected the one that was less costly to manage. In contrast, the portion of the XL air permit that deals with hazardous substances potentially involves hundreds of chemicals, with varying toxicities, exposure routes, and persistence profiles. Furthermore, although exposure guidelines exist for individual substances, the potential risks associated with how exotic substances behave when combined remains poorly understood. Nonetheless, local stakeholders decided, and the EPA concurred, that Intel's binding and voluntary commitments constituted superior environmental performance.

In contrast, non-local environmental groups maintained that scientific uncertainties surrounding chemical trades made objective evaluation of the XL air permit difficult, if not impossible. Rather than rely on a process that compares and trades risks, environmental groups wanted an air permit that encouraged Intel to minimize risk by reducing or eliminate hazardous chemicals. These groups saw XL as an opportunity for the EPA to push firms to achieve continuous improvement through prevention and toxics use reduction.

Indeed, Intel's XL air permit was patterned after a novel permitting experiment at one of its older plants in Aloha, Oregon. In the case of the Aloha effort, the older plant, a major source, was effectively reclassified to escape reporting requirements; furthermore, it received a narrowly defined list of pre-approved process changes under minor source provisions in exchange for enforceable pollution prevention commitments (ODEQ 1994). In contrast, the XL agreement contains no explicit pollution prevention commitments for Fab 12. However, Intel maintains that the only way it could meet the emissions limits and build a second fab at the site would be to incorporate prevention concepts into the product and process blueprints for second fab. As was discussed in chapter 2, Intel reports that its engineers and designers have successfully used "Design for the Environment" to reduce some types of emissions in the past. For example, the new processes installed in Fab 12 emit 40 percent less VOCs than processes for previous chip generations (Mohin 1997, p. 10351). Such environmental strategies are consistent with recent efforts by major chip companies to minimize production uncertainties.

By integrating manufacturing concerns into design, Intel is able to reduce production uncertainties years before the methods are transferred to high-volume fabs. Viewed from this perspective, it is possible that the XL air permit encouraged Intel to better integrate environmental goals into product design—even if a second fab is not built at the Ocotillo site.[1] No

1. The prospect that one will be built there appears increasingly unlikely.

matter where Intel builds its next fab, it is virtually impossible to assess whether or not the XL air permit caused the company to prevent more pollution than it was already planning to prevent, owing to the intense secrecy of Intel's product development. Intel is so concerned with disclosing product development data to competitors that it now uses names of western rivers to cloak the identities of new chips. Thus, it is highly unlikely that Intel would elect to disclose exactly how environmental goals affect the design of its products. Yet, in order to be successful, projects such as XL require companies to disclose more information to the public than traditional regulations require.

Differing Interpretations

Most experts on environmental policy agree that, if a company can control more pollution that regulations require, the EPA should give it the ability to do so (NAPA 1995; Clinton and Gore 1995). In the Intel case, however, agreement stopped there. One reason for this was the lack of firm "baseline" measures and technical information on chemical risks. Closely related to this was the question as to whether prevention or comparative risk trading is the best way to make decisions in the face of uncertainty. Another reason—one that has received less attention—is that, in order to encourage experimentation, the EPA deliberately refrained from defining clearly what Project XL was supposed to achieve. The EPA's decision to leave XL's goals and objectives ambiguous reinforced radically different expectations about what the initiative was supposed to achieve.

Intel and some other firms that considered participating in the initiative were under the impression that Project XL would serve to showcase and reward existing achievements, such as Intel's ability to reduce VOCs by 40 percent relative to previous chip generations. In contrast, environmental groups and environmental-justice organizations saw XL as a way to encourage participating companies to ratchet pollution downward.

The idea that participation in Project XL would require companies that already overcomplied with current regulatory standards to reduce emissions even further struck a number of firms as preposterous. One Intel employee told a Washington research team: "We're being measured against some theoretical nirvana. No matter what we do it's not enough. You have to be Jesus Christ to qualify for XL." (NAPA 1997, p. 86)

Intel saw Project XL as a way to showcase its $28 million wastewater facility and its donations of used computer equipment to community programs. The EPA and other Project XL stakeholders concluded that these voluntary commitments, coupled with Intel's pledges to improve its handling of wastewater, solid waste, monitoring, and reporting, constituted environmental performance superior to what would have occurred in the absence of Project XL. Environmental groups charged that the EPA's decision to credit Intel for what the company arguably could have achieved without XL failed to constitute superior environmental performance.

The EPA also could have helped to minimize controversy by administering Project XL along with the Common Sense Initiative, the agency's parallel reinvention project. As was described in chapter 5, the electronics and computer sector participants in the Common Sense Initiative—including Intel—developed a "Facility-Based Alternative System of Environmental Protection." The vision statement would provide companies regulatory flexibility for facilities that "commit to superior environmental performance" (US EPA 1995c). The statement calls for the development of an environmental management system approach that sets performance objectives and stresses continuous improvement. Perhaps most central to the vision statement is the concept of proportionality: facilities that seek substantial regulatory flexibility must demonstrate equally substantial improvements in environmental, health, and safety performance beyond what current regulatory standards require. In addition to the vision statement, the computer and electronics sector participants—including Intel—launched a CSI pilot project to identify existing regulatory barriers that prevent companies from achieving zero wastewater discharge. However, Project XL failed to build upon (or even to consider) the work already underway in the Common Sense Initiative—a move that outraged environmental and environmental-justice group participants in that initiative.

The Silicon Valley Toxics Coalition, for example, maintained that XL missed an opportunity to test CSI pilot projects. As an illustration, consider that, while CSI participants sought to achieve zero wastewater discharge, Project XL provisions allow Intel to treat wastewater to levels that enable it to be used as drinking water. Though the water meets federal heath standards, it contains too many impurities to be reused in wafer fabrication. From Intel's perspective, the ability to cleanse water used in manufacturing to the point where it can be re-injected into aquifers represents a real achievement. But environmental groups question how it is possible for water that is too dirty for wafers to be fit for human consumption.

The Natural Resources Defense Council and the Campaign for Responsible Technology also criticized the EPA for failing to use the additional flexibility under the XL agreement to persuade Intel to commit to environmental improvements at least equal to the enforceable pollution prevention provisions achieved at Intel's facility in Aloha, Oregon (US EPA 1996f). Unlike the P4 permit in Oregon, Intel's XL air permit contains no commitments for the manufacturer to reduce pollutants over time.

In the case of Fab 12, a minor source, both Intel and the EPA countered that additional pollution prevention requirements requested by environmental groups would give external actors too much control over the XL process. In a column in the 28 June 1996 *Washington Post,* Timothy Mohin, Intel's government affairs manager, asked incredulously: "Citizens are going to make decisions . . . that are binding on Fortune 500 Companies?" However, a number of environmental and environmental-justice groups were under the impression that the EPA had designed XL to give public stakeholders greater influence over companies' environmental decisions.

Who Decides?

In the absence of existing technical standards, initiatives such as XL require participants to negotiate outcomes. Without a legal mandate, those affected by the outcome of the negotiation must have an opportunity to evaluate options and consent to the final outcome in order to avoid the possibility of lawsuits later on.

However, the integrity of the Intel agreement was severely compromised by the EPA's failure to define sufficiently who could participate and how participants would weigh and decide difficult tradeoffs, particularly those involving hundreds of incommensurable chemical risks. David Matusow, one of the XL public stakeholders, observed: "The public entering Project XL cannot be expected to be experts in the technical issues raised by this activity. In fact, they will seldom have this background; this type of knowledge is not within the normal experience of the public." (Matusow 1996, p. 1) Indeed, public participants received no independent technical assistance to help them evaluate the terms of the XL agreement. According to Mohin (1997, p. 10349), regulatory agencies and Intel provided analyses and technical expertise to public stakeholders. However, Matusow said that the lack of outside technical assistance puts the public at a disadvantage in the negotiation effort because both Intel and regulators are not disinterested parties but have a large stake in the outcome.

In addition to lingering questions about how the lack of independent, third-party technical assistance may have colored the outcome, others question the composition of the "public" stakeholder team. Some (see, e.g., Lewis 1997, p. 7) claim that public participants were not drawn at random from the local community but in fact were representatives of institutions with a strong interest in Intel's continued economic success. Among the participants in Project XL were the planning director for the city of Chandler, a local school board member who was also a member of the Chamber of Commerce, and a director of the Arizona Public Health Association.

The participants may have represented organizations with vested interests. However, it is far from clear whether their perceptions of Intel departed significantly from those of the average Chandler resident. By all accounts, Chandler residents view Intel as an outstanding corporate citizen. They credit Intel with transforming what was once thought of as a little town on the road into Tucson to a city that the *Arizona Republic* characterized as "young, smart and well-off" (22 October 1995). Before adding Fab 12, Intel employed 4800 in the region. The new fab called for 2000 additional workers. According to Intel, the good will that the firm had fostered in Chandler made the XL agreement possible.

In addition to serving as an employment source, Intel has sought to improve the community's technical literacy. According to the *Arizona Republic,* Intel has fitted the local high school for computer technology and has donated 3 acres of land for elementary schools. At one elementary school, Intel has sought to improve math and science courses. Intel also has donated $2 million to enable the city library to purchase Intel-based computers.

Aside from the charges that public participants represented organizations with interests in Intel's economic success, some say the selection process was biased in favor of Intel. All five of the participants in the stakeholder process were drawn from a preexisting Intel Community Advisory Panel. Approximately 4 years before XL was announced, Intel established a Community Advisory Panel to solicit input from Chandler residents. Intel has established similar Community Advisory Panels at its other manufacturing sites. Environmental groups have criticized these panels for lacking transparency, controlling media access, and being unresponsive to community concerns (CRT 1996b, p. 6).

Intel representatives counter that the XL model was selected only after an exhaustive assessment of stakeholder participation models (Mohin 1997, p. 10348). For example, Intel reviewed all stakeholder processes classified as models by environmental and environmental-justice organizations participating in the Common Sense Initiative. According to Mohin (ibid.), the Project XL stakeholder process was "a far more extensive and inclusionary program than any of the examples [Intel] could find."

If the Common Sense Initiative is any indicator, it is likely that the outcome of the XL negotiation would have been very different had non-local groups participated in the Intel stakeholder process. Non-local organizations from Silicon Valley and New Mexico would have drawn from a dramatically different set of experiences with Intel and other chip companies when evaluating the proposal's relative merits. For example, the Silicon Valley Toxics Coalition questioned how the EPA could consider as excellent a company that is named as a potentially responsible party at three Superfund sites in Santa Clara. Similarly, Campaign for Responsible Technology members in New Mexico who contributed to *Intel Inside New Mexico*—a stinging indictment of Intel's activities there—were incredulous that the EPA selected the company to participate in the "Excellence and Leadership Process."

In comparison with other the records of other chip firms in Arizona, and with its own record in Silicon Valley, Intel has a comparatively clean record in Arizona. Though Intel has long maintained a presence in Arizona, its operations in that state are dwarfed by those of Motorola, which operates a dozen wafer fabs in Phoenix, Chandler, and Tempe (SEMI 1994) and which is considered by some to be the state's largest industrial polluter. Six Motorola facilities are reported to have contaminated groundwater in the area around Phoenix (EIGNC 1997, p. 47). The EPA has listed three of Motorola's sites on the Superfund list (ibid.).

One exception to Intel's generally good record in Arizona is a spill that occurred in July of 1993 when workers at a site near Chandler overfilled an underground tank meant to hold solvent waste. In 1995 the state found elevated levels of benzene, 1,1-dichloroethane, 1,1-dichloroethene, 1,1,1-trichloroethane, and Freon TF in aquifers surrounding the facility. Representatives of the Maricopa County Environmental Services Department maintain that Fab 12's emissions to the air are far lower than those from other semiconductor companies in the area—lower, indeed, than those from some smaller businesses (EIGNC 1997, p. 62). The Phoenix-area environmental advocacy group Don't Waste Arizona has given Intel high marks (NAPA 1997, p. 18).

Because local perceptions of Intel's performance depart so significantly from those of non-local organizations, it should come as no surprise that locals strongly resented efforts by the Silicon Valley Toxics Coalition, the Campaign for Responsible Technology, and the Natural Resources Defense Council to participate in the stakeholder process. Locals largely believed that national advocacy groups lacked legitimacy to comment on local decisions. The EPA and Intel maintained that the composition of the XL public group was legitimate because it was made up of people most affected by the agreement's outcomes (US EPA 1996f). Absent from the stakeholder process were Fab 12 technicians and residents of new tract homes that had recently sprung up around the Ocotillo site. Intel claimed to support separate environmental oversight programs for plant employees. According to Mohin (1997, p. 10349), Intel reviewed stakeholder participation models that included employees other than environmental managers and models that excluded them. It then decided to solicit employee input through e-mail messages and company publications. Manufacturing employees were invited to address a Project XL stakeholder panel on workplace health and safety.

In addition to soliciting employee input, Intel sought to attract local residents to eight public meetings on the XL plan through announcements in local newspapers, on radio, and on cable television. For two of the eight meetings, Intel distributed about 20,000 flyers in English and Spanish around town. Materials from the public meetings were placed in local libraries and posted on Intel's web site. Since the air permit ultimately was attached to the Final Project Agreement as a separate, enforceable document, Maricopa County also held a public hearing and a 30-day comment period on the draft air permit provisions. Whereas environmental groups say that the XL process drew too little public attention, Intel, the regulators, and some public stakeholders say that the process was far more inclusive than traditional administrative permitting procedures.

Intel's XL effort appears to have offered more opportunities for the public to participate than traditional laws and regulatory mechanisms. As originally envisioned, Project XL was to be an exercise in case-by-case regulation, with negotiations between the EPA and the regulated firm driving the outcome but subject to stakeholder agreements. Though the XL process may have allowed more opportunities for participation, some say that it should have given people more power to shape the project's outcome.

Project XL also was weakened by different expectations and by the lack of clear guidance from the EPA. For the most part, industry approached XL as a process that would provide more opportunities for stakeholder involvement. Environmental groups and environmental-justice organizations envisioned XL as a way to give people a greater say in environmental decision making. In response to their Project XL experiences, Mohin (1997, p. 10354) wrote that it was "a fundamental misconception that the stakeholder process is a decisional rather than advisory process," and Matusow (1996, p. 4) maintained that community interests were not served by the XL process: "If the benefits to the community are not clear and enforceable, the community participants may not only come away feeling isolated, exhausted, and stressed financially, with a greatly reduced sense of trust in the governmental process and industry's claim to care about the community in which they conduct business."

The issue of community and worker participation points to one of the central challenges in moving toward an alternative system of environmental governance: if the United States seeks to adopt an XL-type system of environmental management, who is a legitimate representative of the public's interest? Should public participants in XL-type processes serve as advisors, or have a say in a company's environmental decisions?

At least in the case of the Intel effort—an experiment that involved a multi-national corporation and a set of novel permitting strategies—it is clear that national environmental and environmental-justice groups should have been included in the official stakeholder process. Excluding non-local groups and interested individuals might be a legitimate strategy in cases where the scope and the consequences of a project are confined to a single community, but not in the case of a path-breaking federal environmental experiment involving a multi-national corporation. Non-local environmental and environmental-justice groups sought to participate in Intel's experiment to keep the company from using its enormous influence to co-opt local residents and regulators—something that federal environmental laws were presumably designed to prevent.

Intel's XL agreement fails to violate any federal, state, or local laws and contains provisions of interest to Chandler residents. However, Project XL is a high-profile federal policy experiment, and Intel's XL agreement involves a company with plants in several states and several countries—places and political jurisdictions where "outside" interest groups may legitimately claim stakes in the project's outcome.

Although who participates is important, a thornier and more relevant question is whether public stakeholders should play an advisory or a "decisional" role in environmental management. In theory, at least, if the nation is to move toward a system of case-by-case standard setting that provides equal if not superior environmental performance, the people who might be affected by a project's outcome must be able to negotiate binding commitments from a company. It is clear that public stakeholders will be at a perpetual disadvantage in the decision-making process as long as they fail to possess unbiased information with which to evaluate industry claims.

However well-intentioned the efforts by Intel and regulatory agencies to supply public participants with technical expertise may have been, the fact remains that both Intel and the regulators had vested interests in the project's outcome. Matusow (1996), summarizing his XL experience, wrote that unequal access to technical information "results in the public stakeholders being put into an inferior position at the negotiation table for Project XL," and that "this disadvantage is so large as to all but exclude them from the team."

Beyond participation issues, the question remains as to why the EPA selected a company that represents an industry where "only the paranoid survive" for an experiment whose very success relies on greater public access to information. It may be the case that emissions limits contained in the XL agreement compel Intel to modify its processes and products to meet environment goals. However, it is unlikely that Intel or any other major chip manufacturer would elect to specify publicly how it modifies its plants and equipment to meet such goals. From this perspective, Intel's secrecy puts it in a no-win situation relative to groups seeking assurance that environmental goals are being met.

Who Benefits?

Project XL's *raison d'etre* was to give firms the ability to control pollution in a more efficient and effective manner. Potential cost savings to Project XL participants were diminished from the outset because the legality of the initiative was not certain. As a result, participants in the EPA's initiative—including Intel—developed projects that largely complied with existing federal, state, and local laws. For Intel, though, the XL air permit pre-approval provisions could provide significant benefits by reducing delays.

Permit notification and review processes can impose delays of weeks, months, or even years. In the presence of competition, delays threaten to erode slim technological and marketing leads. In addition to the potential for delays, environmental regulations that require manufacturers to notify regulators each time a production change occurs impose considerable paperwork requirements and incur permitting fees.

Unfortunately, determining the net economic benefit to Intel of the air permit is likely to be difficult, if not impossible (Boyd et al. 1998, p. 50). The difficulty is due in part to the site-specific nature of the XL permit and in part to the complex effects that any form of regulation—including command and control—can have on the private sector. However, this difficulty should not be used as an excuse to avoid experimentation with regulatory flexibility.

Intel, both before Congress and in formal comments to the EPA, has repeatedly made the case that permitting provisions threaten to harm its competitive position (Sheppard 1995). However, in the same way that the EPA failed to develop data that would allow hazardous emissions from Intel's fab to be compared with those from other facilities, neither the veracity of Intel's claim nor the potential economic effects of a permit on Intel's competitors was considered before or during the process.

Determining precisely whether delays due to permitting harm Intel, or whether the XL provisions harm Intel's competitors, requires a level of technical detail beyond the scope of the most sophisticated economic analyses. As was discussed in the introductory chapter, the lack of models is likely to be attributable to the fact that economists tend to avoid problems that they are currently unable to formalize (Krugman 1991, pp. 6–7).

Intel manufactures dozens of Pentium-class products and is usually unrivaled in the markets for the latest such products. In the markets for more mature products and less sophisticated microprocessors, competition is generally stiffer. Therefore, at a minimum, knowing whether Intel benefits from a permit that reduces production delays requires determining how the market for the microprocessors made at Fab 12 is organized.

Intel characterizes the product manufactured at Fab 12 as a "leading-edge Pentium-class product with multimedia capabilities" (Mohin 1996). However, Intel has declined to specify exactly what type of product Fab 12 manufactures. According to *Microprocessor Report,* an industry trade publication, the product manufactured at Fab 12 is most likely the P55C chip with MMX multimedia extensions that enhance video display (Slater 1996a). According to trade reports, Intel's two main competitors in the P55C market are Cyrix's M2 chip and Advanced Micro Devices' K6. Although data on the P55C market are not available, it is known that Intel supplies more than 80 percent of all microprocessors. If Intel's share of the market for the P55C chip is comparable, the threat of losing market share as a result of delays in permitting, though real, is likely to be negligible.

Cyrix, being fabless, does not face the direct threat of delays in permitting. The chief threat to Cyrix is that its foundry partners, IBM and SGS-

Thomson, may fail to fill orders in a timely manner. Advanced Micro Devices, which does fabricate chips, simply lacks Intel's production capacity. Though it is not implausible that permit delays could erode some of Intel's commanding lead, that company's claim that delays due to permitting require it to build new fabs outside the United States is probably overstated.

To evaluate broader economic effects associated with Project XL, it also is necessary to consider the project's administration costs. The US EPA, the state of Arizona, and Maricopa County all had to divert resources from programs required by law. Similarly, Intel and the interested individuals and advocacy groups had to divert time and resources from other activities to XL. Because the XL experiment involved negotiation, it was understood that the initial transaction costs to industry, regulators, and public participants would be high, but it was hoped that the cost reductions accorded by increased compliance flexibility would more than make up for the delays and costs of negotiations. Project XL's architects also hoped that transaction costs would decrease over time as additional facilities and firms adopted XL programs.

The Truth Is Out There

What was originally thought by Intel and the EPA to be a straightforward 6-month exercise instead turned into a protracted and contentious process that required more than 11 months and more resources than any of the participants, including Intel, originally envisioned. The experience has prompted all parties to rethink the administrative feasibility of tailoring regulation to individual facilities. One participant noted in a follow-up report that Project XL was "not a game for the timid" (Coombs 1996, p. 1). Similarly, the EPA noted in a follow-up report on regulatory reinvention that "the resource demands of evaluating and negotiating thousands of site-specific agreements would simply be too great" (EPA 1997a, pp. 1–2) The EPA, furthermore, "does not expect Project XL to provide 'the' definitive answer on the next era of environmental protection being debated" (ibid., p. 2). The agency's desultory tone is a far cry from President Clinton's laudatory remarks on the potential of XL just 11 months before Carol Browner traveled to Chandler to sign the Intel agreement.

Though the parties agree that the process left them feeling short-changed, there is little consensus as to why Project XL fell so far short of expectations. Environmental groups call the initiative a "regulatory free-for-all," according to a 24 January 1997 *Washington Post* article. Others blame national environmental interest groups, such as the Natural Resources Defense Council, for lodging unsolicited technical and legal objections after local XL stakeholders and regulators had already approved the plan. Others believe that the failure to pursue broad-based public participation contributed to the sub-optimal results. In response, the EPA made a number of mid-course corrections to remedy problems identified during the Intel project. It is clear that the controversy surrounding Project XL could have been minimized had the EPA more clearly defined the project's goals, objectives, and operating procedures in advance.

The Aloha permitting project generated less controversy than Project XL because the scope of the projects, the number and complexity of the potential tradeoffs, and the number of participants were relatively constrained, so there were fewer opportunities for the process to go awry. Though it struck a blow against openness and greater public participation, part of the reason that the Aloha permitting effort was considered a success was that public participants and lawyers were excluded from the work group (PNPPC 1995, p. 18).

It is possible that the permitting features of the XL agreement do, in fact, encourage Intel to incorporate environmental concerns into the design process. However, it is clear that Project XL was not the best vehicle for Intel to showcase its environmental achievements. Furthermore, the Oregon permit shows that the slow, contentious XL process probably was not the most cost-effective and expeditious way for Intel to obtain an air permit.

Air permits that approve routine process changes in advance may give a manufacturer more flexibility to select the most efficient abatement methods and may help minimize the prospect of production delays. However, measuring the magnitude of such benefits is complicated by inadequate environmental benchmarks. To know whether the XL permit delivers benefits, it is necessary to know what the emissions from Fab 12 and from a second planned fab would have been in the absence of XL. The lack of data on the composition of the market for the products made at Fab 12 and possibly at a second fab further complicates efforts to evaluate the economic impacts of the agreement on other firms, on downstream industries, and on society. Project XL's benefits to Intel's competitors are still too complex to be estimated confidently (Boyd et al. 1998).

In addition to the lack of adequate participation and accountability mechanisms, XL suffered from an inadequate framework for assessing the industry's locational features. As a result, environmental decisions regarding a multi-state, multi-national corporation were approved by a handful of local residents selected by Intel. In view of their city's excellent economic relationship with Intel, it is not surprising that Chandler residents weighed the merits of Intel's proposal differently than organizations from other parts of the country that have experienced the environment effects of microchip manufacturing firsthand. As was discussed in the previous chapters, Congress expanded federal involvement in environmental laws precisely to discourage states and localities from placing economic concerns above environmental interests. By ignoring Intel's performance in other places, and by excluding as participants public interest groups outside Chandler, Project XL inadvertently appeared to reward localities that fail to consider fully the economic and environmental costs associated with manufacturing microchips.

Silicon Evolution

. . . I'm sure that silicon intelligence is going to evolve eventually to the point where it'll get harder and harder to tell intelligent systems from human beings.
—Gordon Moore, co-founder and chairman emeritus of Intel (Port 1997)

From Model Ts to Microchips

It is difficult to exaggerate the degree to which the semiconductor has transformed and will continue to transform not only the composition of the economy but also the ways people work, govern, and play. In roughly a third of an average human lifetime, the high-technology industry, powered by the microchip, has grown from a handful of "traitorous" engineers to a growth engine that helps drive the US economy and the global economy.

Yet the simple economic models on which environmental policies are premised tend to target pollution that results from the manufacturing of Model Ts. That is, vertically integrated manufacturing models that depict research, design, and manufacturing under one factory roof are increasingly inappropriate now that a chip planned in Silicon Valley may be manufactured in Taiwan, assembled in Singapore, and sold in Puerto Rico. Understanding the environmental effects of an industry in which profitability depends on moving products rapidly to market and production increasingly takes place hundreds if not thousands of miles away from product planning and design poses considerable analytic challenges.

The Distemper of Democracy

In contrast to fluid and fast-paced manufacturing, democratic processes are cautious, cumbersome, deliberative, and often contentious. It is therefore not surprising that microchip manufacturing appears to outstrip government's and civil society's ability to effectively evaluate—much less regulate—its positive and negative environmental impacts. To what degree should social institutions try to match or mirror the industry's tempo and structures? Is chip manufacturing so fundamentally different from other industrial forms that it requires unique laws and regulations? Or does the cautious process of administrative rulemaking serve as an effective instrument to blunt the industry's negative impact on humans and the environment? This book has used the case of the microchip to explore how well laws and regulations to protect human and environmental health intersect with their manufacture.

Wafer fabrication is distinguished from other manufacturing forms such as chemical production or petroleum refining by the degree to which chips' value is a function of continually improving designs ideas and production methods. Whereas a chemical plant's processes may last up to 15 years, manufacturing equipment inside leading-edge chip fabs lasts less than 5 years—enough to make two generations of products. Yet both manufacturing forms commonly rely on process chemicals. Chips are not manufactured in a vacuum but in rarefied environments that rely on some of the most toxic chemicals and gases in contemporary manufacturing. Furthermore, the quantity and the identity of the substances used to make defect-free chips are constantly changing.

Replacing Old Problems with New Ones

To some degree, the emergence of this novel form of manufacturing underscores the importance of traditional laws and regulations designed to minimize risks: technology may solve old problems, but it tends to create new ones. The automobile, which virtually eliminated urban horse manure, gave us urban smog. Far from representing a "clean" new manufacturing method, chip production has presented entirely new policy puzzles.

But the emergence of microchip manufacturing also illustrates some of the present system's fundamental weaknesses. Current environmental laws have helped to stem the flow of pollution to air, water, and land from fabs and other industrial sources, yet the system of federal environmental laws and policies has largely failed to identify and avert problems associated with this emerging industry.

To the extent that rapid change in semiconductor manufacturing makes environmental risks to humans and the environment difficult to identify, rapid innovation and retirement of capital equipment nonetheless represents a tremendous opportunity for manufacturers to reduce the amounts of toxic chemicals, other harmful materials, and water required to make microchips.

As happened with chlorofluorocarbons, glycol ethers, and stepped-up permitting requirements under the Clean Air Act Amendments of 1990, chip and computer firms are consistently among the first to phase out the use of substances. While many observers maintain that environmental laws are ineffectual and that voluntary mechanisms hold greater promise, it is clear that chip firms respond to strong scientific evidence and enforceable measures.

The Dearth of Data

What is most problematic about current environmental laws and regulations is that they give the environmental regulators few effective tools for assessing and targeting the actual environmental impacts of microchip production. For example, according to weight-based and volume-based measures such as the Toxics Release Inventory, the environmental impact of semiconductor manufacturing appears minimal. When viewed in terms of toxicity, however, the materials used to fabricate wafers are among the most hazardous materials now used in manufacturing. Similarly, when emissions from the semiconductor industry are considered as part of the electronics sector, toxic releases and transfers from that sector rank only behind those from the chemicals and metals sector. Continuous chip advances furthermore drive rapid obsolescence of computer and electronic products and consequently contribute to growing hazardous and solid waste. Waste from discarded personal computers may intensify as chip makers and hardware manufacturers develop network-based machines, thus making today's desktop computers obsolete.

Evaluating the environmental impact of chip manufacturing, as well as the degree to which environmental regulations may effect competitiveness of chip makers, has become even more challenging as US firms respond to stepped-up international competition. Now, a number of small US chip companies no longer manufacture chips, but contract production to Asian suppliers. Those firms that continue to manufacture chips increasingly do so through complicated alliances with Japanese, Korean, and German competitors. For Intel and for other companies that are still able to independently bankroll multi-billion-dollar fabs, corporate reorganization makes it possible to locate a new fab almost anywhere.

Defining Boundaries

Organizational and geographic realignment can contribute to misleading impressions about the environmental performance of the chip industry. Fab closure, subcontracting, and the increasing tendency of US chip companies to build new fabs overseas near foreign markets contribute to declines in the industry's toxic releases and transfers. Fluid subcontracting relationships and strategic alignments further blur environmental accountability and minimize opportunities to improve the environmental performance of overseas suppliers.

The ability of US firms to build fabs anywhere gives them enormous leverage over localities eager to lure coveted manufacturing jobs. Some environmental groups and environmental-justice organizations maintain that economic development strategies that ignore Silicon Valley's environmental legacy will likely be doomed to repeat it.

The ability of chip firms to locate production facilities anywhere may lower the cost of doing business, but building fabs in regions that lack an existing microchip manufacturing base can carry hidden costs to firms. Economic geography demonstrates that the enormous success of Silicon Valley chip firms partially stems from the existence of supporting institutions—universities, suppliers, interest groups, trade associations—and from local government policies that recognize the industry's unique needs and problems. Although building new wafer fabs in regions that lack one or more of these elements may offer some advantages, locating fabs in states such as Virginia is accompanied by other issues arising from regulators' relative unfamiliarity with what chipmaking requires (such as an adequate infrastructure for wastewater disposal). Industry relocation also may raise costs for local residents. For now, people who live in regions that lack watchdog groups to monitor firms and disseminate information must independently bear the cost of evaluating whether it is safe to live near a wafer fab.

Legislative Authorization

The policy challenges of chip manufacturing have not escaped notice by the Clinton administration. Rather than revise existing laws, the administration has advanced several initiatives to identify what is wrong and to build consensus as to whether laws and regulations should be updated. Since Bill Clinton assumed office, the US Environmental Protection Agency has unveiled two prominent experiments designed to improve how laws target information technology firms. The Common Sense Initiative is aimed at the computer and electronics sector. The centerpiece of Project XL is an Intel fab in Chandler, Arizona. Both initiatives have failed to provide definitive answers as to whether the statutes need fixing, because both have had their efficacy diminished by the Environmental Protection Agency's lack of legal authority to administer them.

Project XL and the Common Sense Initiative demonstrate that experimental initiatives require legislative backing to reduce legal risks to participants and to provide the EPA with adequate resources to administer such programs. As an illustration of the importance of legislative backing, consider that the sulfur dioxide emissions trading program authorized under the Clean Air Act Amendments of 1990 (Title IV) has been one of the most successful alternative regulatory strategies to date (Burtraw 1996). Unlike Project XL or the Common Sense Initiative, the trading program possesses specific pollution reduction goals, timelines, and, perhaps most important, a strong analytic framework to help regulators choose firms to target. Congress designed Title IV to reduce 10 million tons of sulfur dioxide emissions from 1980 levels by 2010. Also unlike

XL or CSI, the emissions trading program was premised on 30 years of theory and experimental study in environmental economics aimed at examining the efficacy of market-based instruments. On the basis of such efforts, decision makers and environmental managers were able to determine in advance which industries were best suited for emissions trading schemes. As a result, Title IV primarily targets electric utilities and plants with similar pollution and production profiles.

Though legislative authorization of Project XL could have reduced risks to participants and resulted in more daring proposals, a more fundamental question is whether chipmaking requires different regulatory strategies. The chief complaint of chip companies is that regulatory instruments such as permits increase the risk of production delays. In practice, the potential for Intel to experience delays due to air permits at fabs where chips are already in production is minimal. While complaining that major source permitting provisions of the 1990 amendments to the Clean Air Act would harm competitiveness, Intel nonetheless adopted a corporate policy of structuring almost all its US facilities to emit pollutants well under the threshold limits that would trigger reporting for major sources—a significant achievement. One exception is a facility in Aloha, Oregon, that houses some older production lines and is a major source of air pollution. There, Intel worked with the state of Oregon and the EPA to develop a so-called P4 ("pollution prevention in permitting pilot") that places a plant-wide cap on emissions and approves in advances routine production changes at the older fab as well as a second planned facility. In exchange, Intel commits to fully enforceable pollution prevention conditions that require it to describe what types of processes and what types of decision making the facility will use to prevent pollution. It also requires Intel to establish partnership agreements with its suppliers and equipment vendors to minimize both hazardous and volatile organic compounds from raw materials. Finally, the P4 requires Intel to develop a data collection system and employee training program to promote pollution prevention (ODEQ 1994, pp. 11–12).

In contrast, the air permitting portion of Intel's Project XL agreement formed just one part of a comprehensive plan that covers air, water, and waste issues at Intel's Ocotillo site. However, the permit's pre-approval and emissions cap provisions represent the greatest source of regulatory flexibility under the XL plan. Not surprisingly, the air permit was the most hotly contested portion of the package, drawing the ire of more than 130 environmental and environmental-justice organizations. What was originally envisioned by the EPA to be a 6-month process took nearly 11 months to complete. The lengthy and contentious nature of the XL negotiation makes it questionable whether an XL-type process is the most expeditious or inexpensive way for a company to obtain a permit that helps to reduce manufacturing delays. Provisions to approve in advance routine process changes at an Intel plant could have been achieved through the less prominent, less controversial P4 experiment carried out at another Intel facility in Oregon.

Lack of legal statutes and administrative problems explain part of XL's failure, blame also must be pointed at 1600 Pennsylvania Avenue. In fairness to Congress and to the EPA, it is clear that XL was more a victim of political expediency than of poor administration. The EPA, which often requires years to develop a regulation, was directed to design, assemble, and implement an initiative in a few months.

Whatever the myriad factors that contributed to XL's problems, the unfortunate result is that many companies, non-government organizations, and state and local environmental agencies consider the initiative a failure. If strategies to tailor regulations to industrial facilities are to gain credibility and provide information to illustrate whether laws must be modified, initiatives must be shielded from the four-year political cycle. On 30 October 1997, Senator Joseph Lieberman introduced a bill to provide the EPA the ability to grant facilities and manufacturing sectors flexibility for XL-type initiatives. The Innovative Environmental Strategies Act of 1997 (Senate Bill 1348) would provide the EPA with the resources and the regulatory muscle to carry out such a major undertaking. However, many environmental, labor, and environmental-justice groups say the proposed legislation, like Project XL, is premature. Before Congress gives the EPA the ability to grant companies flexibility, the EPA must develop sufficient participatory and accountability provisions to ensure that case-by-case regulation provides an equal, if not superior, level of protection to humans and to the environment.

Accountability Mechanisms

Even if Congress passes legislation to authorize XL-type projects, it remains unclear whether chip companies make the best candidates for participation. By all accounts, Intel's XL negotiation lacked mechanisms to assure the public that emissions under the novel air permit would provide superior environmental benefits. Because the air permit covers a new facility and allows Intel to construct another fab, it was not possible to develop a baseline against which to monitor whether XL ultimately reduces operating costs and pollution more than extant laws and regulations.

In general, the lack of scientific and emissions data plagues all aspects of environmental management and decision making, not just those associated with chip firms (Davies and Mazurek 1997, p. 16; NAPA 1995, p. 166). For example, few data exist for managers to determine whether the Clean Water Act has really been successful.

However, the data deficit is even more pronounced for chip firms. For example, when the EPA's Project XL staff tried to find data against which to compare air emissions from Fab 12 with similar facilities, they were surprised to find that the EPA had not developed an emissions profile for hazardous air pollutants for US chip companies (US EPA 1996f). Thus, it was difficult to use industry-wide data to determine whether the air emissions from the Intel plant were lower than those from other plants in the industry.

In the same way that the EPA lacked data to determine whether Intel's environmental performance was superior to that of other firms, the agency failed to consider the potential economic effects of the XL project on Intel's competitors. Finally, although the project was designed to be transferable to other facilities, Intel's unique position in the market makes it unlikely that the experiment could be successfully replicated at fabs operated by the company's competitors, who control a much smaller share of the microprocessor market than Intel.

Though appealing in theory, successfully making policies that address industry dynamism requires regulators to develop far more information about environmental and economic impacts of a regulatory experiment than traditional environmental regulations. Because the value to some microchip makers lies chiefly in knowing how to successfully design and manufacture semiconductors, it is clearly not in a company's interest to divulge sensitive information to regulators or to interested public stakeholders. Legal theory and empirical studies have both demonstrated just how difficult it is to disentangle what information indeed constitutes confidential business data. At least one major study with firm-level data found that the industry's confidentiality claims were egregious (Ferguson 1992). If environmental regulators fail to first develop information that better illustrates which firms (if any) merit permitting and reporting concessions, environmental initiatives designed to promote the competitiveness of the US chip makers will merely benefit a few prominent producers.

Analytic Tools

The Intel case shows that the EPA must have a set of selection tools to determine which companies make good candidates for XL-type programs. The agency must have a way of determining whether the benefits of participation to firms and to society outweigh the environmental and economic costs. Efforts are just getting underway to develop analytic models and empirical evidence to help environmental managers choose firms to target (Krupnick et al. 1998). Ultimately, the EPA may discover that facility-based strategies are either poorly or perfectly suited to achieving better environmental results more cheaply at chip fabs.

Although an analytic framework will help the EPA better select participants for tailored strategies, it is unlikely that environmental and environmental-justice groups will support experiments designed to deliver only economic benefits. In addition to reducing costs, tailored regulatory strategies will not gain acceptance among environmental and equity groups unless it can be demonstrated that such strategies do not shift risk to disproportionately impact low-income and minority communities.

Finding Common Ground

Many of Project XL's shortcomings could have been addressed through a parallel EPA effort. Targeted at sectors instead of individual facilities, the Common Sense Initiative provides a framework for industry, regulatory agencies, and public stakeholders to consider jointly whether existing laws and regulations should be modified to reduce costs and pollution. Like Project XL, the Common Sense Initiative was plagued by overoptimistic assumptions regarding the time and resources required to persuade groups that have for years acted as adversaries to find common ground. Nonetheless, the initiative produced a few notable projects that could have helped reduce the controversy surrounding Intel's XL effort.

The Common Sense Initiative's sector team charged with examining regulatory barriers in the computer and electronics industries did, in fact, develop a template for statutory reform that was jointly approved by environmental groups, environmental-justice organizations, and companies (including Intel). The "Facility-Based Alternative System of Environmental Protection" is a vision statement to provide companies regulatory flexibility for facilities that "commit to superior environmental performance" (US EPA 1995c). The document was nonetheless overshadowed by Project XL and by subsequent proposed legislation to authorize XL-type projects. The Common Sense Initiative's facility-based principles could have served as a guideline to govern the experiment at Intel's Ocotillo facility in Chandler.

The Common Sense Initiative could have also served as a mechanism to collect much-needed industry benchmark data to evaluate whether proposals such as Intel's constitute "superior" environmental performance. Ostensibly, data collected through the Common Sense Initiative would be aggregated at the industry or the sector level and thus would be less sensitive than facility-level information under XL. The EPA could have charged CSI participants with the task of developing a jointly approved set of principles to identify what constitutes "state-of-the-art" environmental performance, thus developing much-needed benchmark data for XL.

The lack of agreement as to what should constitute appropriate benchmark data for facilities underscores the challenges the EPA will face should it seek to promote more facility-level projects such as Intel's. In the case of Intel's Fab 12, performance would ideally be measured against a facility that produces identical chips, rather than by the EPA's method, which considered TRI data for a broad swath of firms that report emissions using SIC 3674 as a primary code. The Common Sense Initiative also could have served as a forum in which Cyrix, Advanced Micro Devices, other semiconductor companies might have addressed competitiveness issues resulting from Intel's participation in Project XL.

In addition to gathering environmental data, the Common Sense Initiative, or a similar successor, could serve as a mechanism to evaluate whether SIC codes accurately reflect the organization and geography of industries such as semiconductors. Such efforts may reveal that the EPA's resources would better be targeted at environmental problems that plague smaller chip companies, or at environmental problems surrounding US manufacturers' increased reliance on facilities overseas.

The Common Sense Initiative has not tackled tough questions such as "what is a sector?" or "what is an industry?" However, participants have initiated several pilot projects to explore confidentiality issues. In the electronics and computer sector, for example, pilot projects underway in Texas may help to identify what type of data the public have a right to know and use as well as what data firms may legitimately flag as confidential. In addition to supporting facility-based efforts such as Project XL, findings yielded by the CSI could supplement the Toxics Release Inventory.

Improving the TRI

As chapters 3 and 4 demonstrate, the EPA must modify the Toxics Release Inventory to better reflect how industrial restructuring and globalization affect emissions. IBM recently adopted a policy of using the TRI framework to voluntarily report emissions and transfers from its non-US facilities. Furthermore, a number of countries have adopted similar, but not identical, "public right to know" inventory systems. In recent years the EPA has expanded the number of firms required to report to the TRI and the range of chemicals on which companies are required to report.

Recent revisions of the Standard Industrial Codes may further help to improve the inventory's accuracy but will not account for emissions of US chip makers overseas. The EPA should require US-based companies to report on emissions from all their subsidiaries and facilities around the world. In the interim, the EPA should encourage other major US microchip firms to voluntarily adopt the example set by IBM. In the United

States, the EPA should continue to harness the power of information technology enabled by chip firms to improve data collection and reporting. For example, there is a two-year lag between the time when firms file reports and when the EPA actually publishes public TRI reports. Companies such as IBM, however, now update annually on their web sites TRI data on emissions and transfers. In 1997 the EPA made it possible for firms to file TRI reports electronically. Such advances help to make the reports more timely.

Automation makes it easier for firms to file TRI reports and may help to reduce the time required by EPA to make the reports available to the public. However, the underlying data are based on estimates rather than on actual measures of what toxic chemicals facilities released and transferred. For years, environmental managers at all levels of government have envisioned placing meters on pollution sources to continuously measure emissions. Historically, chip manufacturers have resisted the idea of putting materials meters on individual tools for fear that such gauges might introduce contamination (MCC 1993, p. 108). The measurement methods spelled out under the Ocotillo agreement are consistent with mass-balance estimation methods still used at most fabs.

In addition to technical and economic obstacles, firms may be reluctant to adopt continuous monitoring techniques because such systems threaten to impinge on privacy. Though challenging, such barriers are not insurmountable, as recent industry efforts to ensure consumer credit privacy on the Internet illustrate.

The EPA's Intel effort missed an opportunity to demonstrate the potential power of information technology to protect human health and the environment. As was discussed in chapter 6, accountability was a central source of controversy at Intel's Ocotillo site. Public interest groups and some interested residents feared that the XL agreement would allow the facility to release large spikes of toxic substances and to let the mix of hazardous emissions become more toxic over time. Some residents and national environmental groups unsuccessfully asked Intel to adopt continuous, real-time reporting methods to monitor emissions. Continuous monitoring, rather than the emissions estimates required under the XL permit, could have been used to provide greater assurance that Intel was meeting environmental goals. Instead of showcasing the power of Intel's

Pentium chips to instantaneously convey sophisticated data, the agreement requires Intel to prepare quarterly emissions reports based primarily on engineering estimates.

Short of requiring participants to adopt superior environmental monitoring systems, future XL-type efforts, at a minimum, should require the use of independent auditors to inspect and verify that participants are meeting environmental goals.

Finally, the EPA must be clearer about the goals and objectives of initiatives such as XL. From the industry's perspective, it was understood that the XL initiative was designed to recognize extant achievements. In contrast, environmental groups sought even greater pollution reduction in exchange for increased flexibility. It will take years, and possibly an overhaul of existing laws, to bring these opposing sides closer together. In the interim, the EPA must do a better job of communicating in advance what its high-profile initiatives are expected to accomplish.

Foreign Affairs

Harnessing tools developed by companies whose products run microchips and related transistor devices may eventually help environmental managers to better understand whether microchip manufacturing requires more stringent or more relaxed regulations. However, it will require more than technological refinements to assess the environmental impact of fabless firms and their foundry partners, many of which are found in developing countries.

The environmental policy challenge in developing countries is not so much a lack of data as a lack of resources and political will to administer and enforce environmental laws. Increased reliance on strategic alliances, subcontracting relationships, and informal partnerships are not unique to semiconductor firms, but are perhaps the most pronounced example of how pollution problems increasingly require global management strategies. Internationally, industry and nonprofit organizations appear to be ahead of federal agencies in flagging, documenting, and remedying effects of high-tech manufacturing. For example, since 1994 the industry has sponsored an annual "global summit" on environmental, health, and safety issues. The annual International Environment, Safety & Health

Conferences of the Semiconductor Industry focus on issues such as vendor responsibility, wastewater, and European Union Legislation.

Nonprofit groups are beginning to develop a more international focus. The Campaign for Responsible Technology, for example, has formed ties with Asian, European, and Latin American environmental and labor organizations to better track the environmental effects of globalization.

As the industry increasingly locates manufacturing facilities away from established chipmaking clusters, it is possible that environmental groups must rethink their strategies. When chip firms had no choice but to keep manufacturing close to research and design functions, strategies such as information campaigns and protests were largely effective. The industry's move away from established clusters allows companies to seek out places that are less familiar with the downsides of high tech and which may consequently fail to adequately balance economic and environmental concerns. In such places, confrontational environmental group approaches may be less successful.

The Silicon Valley Toxics Coalition recently published an article that candidly assesses confrontational models used by environmental groups in the United States (Zhang 1997). The author, Jing Yang Zhang—an SVTC student intern from Beijing—applauds the degree to which US environmental groups have harnessed fundamental tools of democracy such as information dissemination and public protests to reduce corporate pollution. However, Zhang questions whether such approaches will work in China, where "conformity is considered more important than individualism." Zhang advises US environmental groups that seek to work abroad to explore alternatives to confrontational strategies: "If environmental groups like SVTC are to foster partnerships elsewhere and work towards sustainability on this planet, they should understand and adapt their strategies to the systems in their counterparts' countries. In many cultures in Asia, backup from governmental authorities is the prerequisite to any substantial achievement."

As an example of an initiative that fosters partnerships, Zhang cites the work of the SVTC and others to reduce the industry's water use through the EPA's Design for the Environment Program. In addition to SVTC, the Intel Corporation and the Digital Equipment Corporation have been extensively involved in the initiative. In contrast to the policy focus of Project XL and the Common Sense Initiative, the goals of Design for the Environment are largely technical. Industry, regulators, and interested groups work together to assess how product and process design can help minimize environmental impacts.

Because US agencies obviously lack jurisdiction over environmental problems abroad, building up the enforcement capacity of regulatory agencies and nonprofit organizations in developing countries may help to minimize the industry's environmental impact. The EPA's Office of International Activities could promote such capacity-building efforts though the Department of State, the World Bank, the Organization for Economic Cooperation and Development, and the Agency for International Development. Because the EPA's statutes requires it to focus on pollution in the United States and its territories, those affiliates can help to better explore the environmental challenges of globalization in the semiconductor and electronics industries.

Devolution Revolution

While some environmental policy challenges are increasingly international in scope, others reinforce the rising prominence of state and local jurisdictions. Intel's XL effort illustrates the degree to which localities increasingly seek to target resources toward environmental issues that interest local residents rather than toward those that interest regulators in Washington. Intel's pledges to improve its management of water and of solid waste—issues of concern to Chandler residents—were sufficient for local residents to grant the company additional air permitting flexibility. Nonprofit organizations from outside Chandler maintained that local residents lacked enough technical expertise to effectively evaluate the effects of Intel's Project XL agreement on human health and the environment. It is therefore not surprising that environmental groups and industry watchdog associations from other parts of the United States found Intel's Project XL agreement inadequate.

In 1997, to demonstrate their frustration with what state officials perceived to be inflexible federal requirements, several states drafted legislation that would allow them to waive most federal environmental rules (Skrzycki 1997). It is tempting but naive to believe that environmental concerns have advanced to a point where all jurisdictions across the United States and its territories possess the resources, the capacity, and the political will to balance economic and environmental concerns. The reality may be that the jurisdictions most eager to lure tax revenues from microchip manufacturing may possess the least resources to ensure that environmental goals are met.

The industry's expansion into regions unfamiliar with the downsides of high technology may appear to offer some benefits in the form of relaxed regulations, but it may also carry less obvious costs. The case in which regional authorities withheld wastewater permits until Dominion modified manufacturing processes at its fab in Manassas, Virginia, illustrates how the lack of up-front information on the industry's unique environmental challenges can lead to costly requirements and potentially increase the likelihood of manufacturing uncertainties. The lack of regional institutions to monitor high-tech firms also could carry costs for people who live and work near wafer plants. In order to learn more about the potential risks of living near a wafer fab, at least one concerned Manassas resident has independently borne the cost of collecting and evaluating environmental information about the industry supplied by groups as far away as Silicon Valley.

The EPA is capable of meeting some but not all of the aforementioned challenges. That agency should begin a process to differentiate between environmental issues that require federal involvement and those that do not. Questions as to whether non-local groups were legitimate stakeholders in Intel's Chandler effort could have been answered if the EPA had made use of economic geography.

In 1995, the National Academy of Public Administration released a report to Congress that explored several thorny geographic questions. Prepared by an independent panel, the report directs the EPA to continue to work on pollution that crosses state and national boundaries, and to prevent the resumption of "smokestack chasing"— competition among states for polluting industries (NAPA 1995, p. 116). In exchange, the NAPA suggested, the EPA should resist the temptation to become entangled in issues that are truly local in nature and in those over which it has no legitimate jurisdiction, such as wildlife management. To help state and local governments improve their capacity to manage pollution problems, the report directs the EPA to "create or maintain incentives that will inspire communities to address difficult problems" (ibid., p. 116). A follow-on study to the NAPA report (NAPA 1997) similarly concludes that the EPA erred by excluding non-local environmental groups from the Intel XL effort.

The Intel effort illustrates just how much work remains to be done after Congress passes laws to give firms and localities more ability to tailor regulations to their needs. The Common Sense Initiative and Project XL suffered from shallow assumptions regarding the time, the effort, and the decision tools required to build trust and to balance the competing interests of groups that have for years acted as adversaries.

Moving away from pre-defined pollution standards also requires competing groups with different perspectives to assess, weigh, and trade risks about which the underlying data remain highly uncertain. Absent concrete information, decisions will be largely subjective, colored by the values, beliefs, and perceptions of those at the bargaining table. The EPA assumed that it was possible for the XL agreement to lead to "win-win" outcomes, with all stakeholders benefiting equally. As the Intel case demonstrates, it is more likely that the process of moving away from predetermined technical standards to negotiated standards tailored to individual facilities creates both potential winners and potential losers.

The EPA must develop a set of protocols and procedures to decide what other actors have a legitimate stake in the outcome, what information and resources they require to make decisions, and how tradeoffs among different options will be weighed and decided. For example, is an industry-watchdog group from one state a legitimate player in a negotiation at a facility in a neighboring state? Will line workers who seek to become stakeholders be shielded from reprisals by their company? Will participants operate by consensus, or by majority rule? What methods can people use to weigh tradeoffs among chemicals with incommensurable risks? As envisioned, XL would have reduced federal oversight in exchange for greater public scrutiny. In practice, XL gave people a greater opportunity to participate, but not the ability to influence project outcomes. In exchange for reduced federal oversight of company actions, will public stakeholders have the authority to "make decisions that are binding on Fortune 500 companies" (T. Mohin, *Washington Post,* 28 June 1996)?

In response to XL's early deficiencies, the EPA has refined the guidelines it uses to determine who may participate and how success will be measured. The EPA also has made available grants to help communities better understand the often arcane scientific and technical details that inform environmental policy decisions (*Inside EPA* 1996a,b; *State Environmental Monitor* 1997, pp. 6–7). Such efforts will help to guide future facility-specific experiments. But care must be taken that new laws and programs are not put into place prematurely. Otherwise, strategies designed to give facilities and localities greater authority could simply result in a throwback to the days in which states set themselves up as pollution havens as a way to compete for industry dollars.

Better Communication of Risks

In almost every case examined in the previous chapters, the controversies surrounding wafer fabs could have been reduced if better information about human and environmental risks had been available. The litigation surrounding IBM's East Fishkill facility and San Jose plastics lab and the adverse publicity surrounding Intel's plants in New Mexico and Arizona stemmed primarily from a lack of means to make certain that chemicals actually are safe. It is likely that Dominion's warm reception in Manassas is due in part to the plant's high degree of openness—something that is still exceptional in the industry.

Although the EPA office that regulates toxic chemicals has collected health and safety studies that document the effects of glycol ethers on miscarriages on female fab workers, the industry has not supported epidemiological studies or risk assessments to determine whether chemicals used in fabrication cause cancer or birth defects. The Semiconductor Industry Association maintains that such studies are unnecessary because until the lawsuit was filed there had been no reports of worker problems. The IBM workers' lawsuit marks the first time that a large group of fab technicians have publicly alleged that their cancers could be tied to chip-making chemicals. Under the current system of laws, it is likely that further studies will come about only when more such worker problems are reported.

The intense secrecy surrounding this industry only raises suspicion and reduces trust among environmental and industry-watchdog groups. If semiconductor manufacturing is indeed cleaner than other industries— as trade organizations and individual companies claim—then it is in the industry's interest to do a better job of making its case to the public. Companies at the vanguard of the information technology revolution must improve their delivery of timely, accurate information on the health and environmental effects of industry processes and practices.

Culture Clash

Better risk communication will likely reduce some of the controversy and some of the hidden costs associated with the industry's expansion. However, Intel's experience in New Mexico and (to a lesser extent) Dominion's balance between Virginia's past and future forms of production point to thornier problems in how the public responds to technological change. Economic history is replete with examples of how people have attacked technology in response to larger social reorganization that seemed out of their immediate control. The best-known example of people taking out their frustration on technology is the Luddite uprising of the English textile workers. It is not unreasonable to assume that some of the rancor stirred up by the microchip industry stems from the public's reaction to broad social changes to which microchips have contributed.

Some may seek to dismiss public fears surrounding the environmental effects of microchip manufacturing as simply a fear of change. But those who would posit that current environmental controversies surrounding microchip manufacturing are an updated form of machine bashing must find a way to reconcile such theories with the hard facts about the industry's dark environmental past. What is more likely is that the expansion of the industry arouses other fears about wholesale changes to the land that the influx of chip and electronics giants threaten to bring about. Intel's experience in New Mexico illustrates how the industry's enormous demand for water is linked by some people to larger forces that threaten to alter their physical environments irrevocably. As one concerned

Corrales resident said: "If the water right is severed, all you can grow are subdivisions and fast food joints." (SWOP 1995, p. 56)

The experience of Silicon Valley demonstrates that such fears are warranted. Indeed, though Silicon Valley currently drives 45 percent of the United States' industrial growth, gone are the fruit orchards, affordable homes, and groundwater supplies that once led residents to call the region the "Valley of Heart's Delight." To some, Silicon Valley's current state is simply the price of "progress." Yet the companies that currently dazzle the world with their latest inventions would do well to consider an admonition penned after the previous industrial revolution: "Men and women who see their livelihood taken from them or threatened by some new invention, can hardly be expected to grow enthusiastic over the public benefits of inventive genius." (Hammond and Hammond 1919)

Though it is generally not in the interest of a company to consider the broader social costs of its inventions and activities, Intel's embarrassing and potentially costly experience in New Mexico suggests that companies—like environmental groups—should take into account the political and cultural norms of the places in which they seek to locate.

Repealing Moore's Law

As chip makers begin to confront the possibility of physical limits to the number of transistors that can be effectively placed on silicon, industry and academia are exploring methods to develop alternative methods of making switches. Some of the new technologies are only minor departures from current manufacturing methods—for example, using alternate substrate materials, such as gallium arsenide. But others represent radical departures. Within 20 years, chips may no longer be manufactured in billion-dollar fabs; they may be grown from self assembling organic molecules or made with microscopic "rubber stamps" rather than costly photolithographic equipment. New chipmaking technologies may eliminate the need for process chemicals. They invariably will create new challenges, as the legacy of semiconductor manufacturing in Silicon Valley shows.

Because it is so difficult to anticipate risks to humans and to the environment, a small but growing number of people in industry, government, and academia seek to prevent rather than to control pollution. Prevention-based approaches encourage industry to design products that reduce the use of toxic and hazardous substances and the demand for raw materials. Pollution prevention also may be introduced into existing manufacturing processes. In contrast to current pollution laws, which treat a factory as a "black box," prevention-based approaches require environmental managers and regulators to possess a thorough understanding of product and process design and manufacturing methods.

Prevention-based approaches are intuitively appealing, but they are difficult to implement for two reasons: First, prevention conflicts with the control-based focus of air, water, and waste laws. Second, it is has not been unequivocally demonstrated that prevention is always effective. The experience of the semiconductor industry shows that there are both promises and pitfalls to prevention approaches. Primarily, it appears that chip firms are more receptive to prevention-based approaches upstream (i.e., during planning and design) than on the factory floor.

Microchip companies may be more receptive to changing the designs of their products than to modifying products already in production, because in production the focus is on improving yield. Strategies that focus on reducing the use of toxic substances in existing manufacturing processes are perceived by such firms to increase the likelihood of contamination, or to reduce yield. The extreme secrecy associated with manufacturing microchips makes it difficult to assess and verify independently the extent to which toxics use reduction approaches would indeed adversely affect yields. However, several of the studies examined in chapter 2 suggest that firms prefer to use a mix of strategies, such as recycling and reuse, rather than to rely on reduction methods alone.

Strategies that encourage chip companies to incorporate environmental criteria in product and process design may be more successful because they are consistent with the industry's focus on long-range product and plant planning. Most of the major chip companies that continue to manufacture wafers have, in recent years, shifted to an organizational structure intended to shorten the time between product design and market sales. To minimize the threat of manufacturing delays, Intel has facilities in which prototype manufacturing processes are perfected months and sometimes years in advance of being transferred to high-volume manufacturing facilities. Intel's fabs, too, are planned years in advance of operation. Indeed, Intel maintains that only through advanced planning and design was it able to commit to emissions limits below major source levels for not one but possibly two wafer fabs under the XL air permit. At the Ocotillo site, Intel may add a second facility that will produce products several generations more advanced than those now being produced in Fab 12. Intel representatives say that the XL permit pushes the company to adopt more stringent product and process design criteria to meet emissions limits for the proposed second facility.

Despite Intel's claims that it is using environmental criteria in product and process design, environmental groups have criticized the XL permit for lacking prevention provisions comparable to those adopted by Intel at its older fab in Aloha, Oregon. But, in contrast with the Oregon case, the XL agreement involves a new facility and a facility still in the planning stages. Intel guards information about its plans jealously, and it considers information about products under development to be so sensitive that it uses code names (such as the names of major western rivers) for them. Such secrecy put Intel in a no-win situation because it is unlikely that the company would ever elect to disclose publicly the environmental considerations it uses when designing products and processes. This underscores the question of whether companies whose products' value is largely a function of information are good candidates for XL-type strategies. Despite these challenges however, the EPA should continue to work with Intel and similar companies on exploring how the permitting process can be modified to encourage the companies to incorporate environmental criteria in advanced manufacturing rather than to try to control problems on the factory floor.

Anticipating Technological Change

Encouraging companies to incorporate environmental concerns into the production of their next-generation chips is important. But even greater priority must be given to research that identifies health and environmental risks at the industry's innovation frontier. Already, there are clear signs that the microchip industry is on the brink of changes as profound as those that were ushered in by the introduction of the personal computer.

Consider that in 1997 three major US chip companies—Intel, National Semiconductor, and Motorola—announced plans to develop new, low-cost network-based devices that would make desktop personal computers obsolete. Indeed, Motorola scrapped its original designs for a microprocessor fab near Richmond, Virginia, and now plans to devote part of the proposed plant to research and development work on non-silicon-based chips.

The adoption of new substrate materials may "solve" some of the environmental problems associated with manufacturing silicon-based chips. The transition away from silicon-based chips may reduce the use of water and chemical solvents. Manufacturers' attempts to reduce contamination by using robots and by keeping wafers contained throughout the manufacturing process may already be reducing workers' exposure to toxic chemicals. Automation also may reduce the industry's demand for production-line workers.

History suggests that innovation is likely to create new environmental challenges. For example, gallium arsenide, which is likely to replace silicon in some applications, is a derivative of arsenic, which is poisonous to humans. Depending on the type and the level of exposure, arsenic can result in skin lesions and in skin, kidney, lung, and liver cancer. Under the current system of environmental laws, however, problems associated with new technologies will be flagged only after they are created. Some pollution prevention principles are currently being applied to computer and electronics manufacturing through the efforts of the public-private research consortium Sematech and through federal initiatives such as the EPA's Design for the Environment.

The pending technological shift also will affect the hardware and software industries. For example, manufacturing network-based computers may require smaller amounts of materials than manufacturing today's personal computers, but the transition to network computers also may intensify waste problems by making today's desktop machines appear unwieldy and obsolete.

Federal environmental initiatives such as the Common Sense Initiative and Design for the Environment are incremental mechanisms for improving how the current legal system anticipates future environmental and human-health risks. However, shifting toward an approach that averts rather than controls air and water pollution and waste problems requires new environmental legislation. As the controversies surrounding the goals and objectives of Project XL and the Common Sense Initiative illustrate, such major changes in the laws will be difficult to achieve. There is currently no consensus on what is broken or how to fix it. Despite their shortcomings, though, both Project XL and the Common Sense Initiative have the potential to demonstrate what is wrong and how it should be remedied. In order to better administer these projects, or their successors, Congress should give the EPA the authority and the resources to administer sectoral and facility-based initiatives. But Congress should not adopt such laws before certain thorny problems regarding participation, decision-making, and accountability are resolved. If experiments in environmental policy are to yield meaningful information, there will have to be honest and objective evaluation of programs and projects—preferably by federal agencies other than the EPA or by nonprofit organizations.

Economics and the Environment: Why the Twain Should Meet

Aside from the environmental challenges they pose, the physical limits on the number of transistors that can be successfully etched in silicon raise questions about economic development policies intended to lure fabs to certain states or localities. Economic developers and planners are starting to understand that thriving "chip clusters" require research and design functions, a supplier network, and supporting institutions. Virginia economic developers demonstrated a solid grasp of these principles when they encouraged Motorola not only to build a large fab near Richmond but also to house chip research and design functions there. When measured against other economic development strategies, such as giving $200 million sports stadiums to billionaire team owners, development strategies based on microchip manufacturing look pretty good. Similarly, federal environmental policies to encourage chip companies to continue to build manufacturing facilities in the United States, perhaps creating 40,000 domestic manufacturing jobs, may be preferable to trade policies that some say send manufacturing jobs overseas.

Economic geography and the related discipline of urban and regional planning have not yet incorporated a perspective that balances development and environmental concerns. Consider the case of Virginia. To ensure that chip manufacturers create manufacturing jobs, planners have prudently tied the payoff of incentive packages to whether or not companies fulfill their long-run expansion plans. From an economic development perspective, the strategy is commendable. However, state and local officials failed to concomitantly examine whether sufficient infrastructure was in place to handle an enormous amount of wastewater. Dominion Semiconductor is sure to lose credibility in Virginia if people there come to believe that the planners failed to balance economic and environmental concerns.

Even more pressing is the question of what will happen to Virginia's three new plants 20 years from now, when state incentives expire and microchip manufacturing methods are likely to have changed dramatically. The strict liability provisions of federal laws such as those associated with the Superfund may discourage further contamination such as that caused by chip and electronics companies in Silicon Valley. But the Superfund law, like most federal environmental laws, is administered by the state—a state that in recent years has been sharply criticized by the EPA and by environmentalists for its weak record of enforcement against industrial polluters. In case the three plants are not in operation 20 years from now, what planning measures are being taken to ensure that today's industrial greenfields will not be tomorrow's contaminated brownfields?

Creating high-paying manufacturing jobs is commendable. But economic success, as the case of Silicon Valley shows, is often achieved at high costs to human health and to the environment. Ted Smith, director of the Silicon Valley Toxics Coalition, has observed that "in fact, we have all learned that the cost of doing business badly—measured both in dollars and in reputation—has become prohibitive." (Smith 1992)

In the same way that economic geography and planning must incorporate environmental concerns, prevention-based approaches should consider insights from industrial organization and geography. Because they do not do so, current prevention approaches lack a framework in which to address the environmental challenges associated with fablessness and those associated with the development of advanced manufacturing facilities that allow large companies to build new fabs far from established chipmaking regions. Environmental managers will continue to miss the moving target of rapidly changing semiconductor manufacturing as long as they fail to view environmental policy from a perspective that takes structural change into account.

In view of the importance of the semiconductor industry to the US economy and to the global economy, industry, regulators, and interested public participants must continue to jointly explore ways to build cleaner, faster microchips. Building better microchips requires cooperation between groups that have traditionally acted as adversaries—cooperation that will not happen overnight, or over the course of any presidential administration. Cooperation requires time, resources, and a common vernacular. By illustrating how insights from economic geography can strengthen environmental policy, and vice versa, this book provides a lexicon that might inform efforts by government, regulators, industry, and environmental groups to work together to create environmental programs based on principles from microchip production rather than from "Model T" production.

References

Allenby, Braden R., and A. Fullerton. 1992. Design for Environment: A new strategy for environmental management. *Pollution Prevention Review,* winter: 51–61.

AMD (Advanced Micro Devices). 1998. Manufacturing facilities overview. Available at http://www.amd.com.

Amoco and US Environmental Protection Agency. 1992. Project Summary, Pollution Prevention project (June).

Angel, David P. 1994. *Restructuring for Innovation: The Remaking of the US Semiconductor Industry.* Guilford.

Arrigo, Linda Gail, Tze Luen, and Yvonne Lin. 1996. Environmental Conditions and Environmental Law in Taiwan. Available at http://taiwanese.com.

Ayres, Ian, and John Braithwaite. 1992. *Responsive Regulation: Transcending the Deregulation Debate.* Oxford University Press.

Barbera, Anthony J., and Virginia McConnell. 1990. The impact of environmental regulations on industry productivity: Direct and indirect effects. *Journal of Environmental and Ecological Management* 18, no. 1: 50–65.

Barlas, Pete. 1997. Amkor Tech seeks funding for entry into wafer fab market. *Silicon Valley Business Journal,* November: 5.

Behr, Peter. 1997. Neighbor on a high tech hunt: A financial service slump pushes Delaware to pursue semiconductor jobs. *Washington Post,* 31 March.

Bernardini, Oliviero, and Riccardo Galli. 1993. Dematerialization: Long-term trends in the intensity of use of materials and energy. *Futures* 25, no. 4: 431–448.

Bérubé, Michael R. 1992. Integrating Environment into Business Management: A Study of Supplier Relationships in the Computer Industry. MIT Technology, Business and Environment Research Group.

Blum, Justin. 1997. Chip plant considers 26-Mile Va. pipeline. *Washington Post,* 11 July.

Borrus, Michael G. 1988. *Competing for Control: America's Stake in Microelectronics.* Ballinger.

Borrus, Michael G., J. Millstein, and J. Zysman. 1983. Trade and development in the semiconductor industry: Japanese challenge and American response. In *American Industry and International Competition,* ed. J. Zysman and L. Tyson. Cornell University Press.

Boyd, James, Alan J. Krupnick, and Janice Mazurek. 1998. Intel's XL Permit: A framework for evaluation. Discussion paper 98–11, Resources for the Future.

Bradley, David. 1996. Tiny rubber stamp succeeds where chip makers fail. *New Scientist,* 29 June: 18.

Brennan, Timothy J., Karen L. Palmer, Raymond J. Kopp, Alan J. Krupnick, Vito Stagliano, and Dallas Burtraw. 1996. *A Shock to the System: Restructuring America's Electricity Industry.* Resources for the Future.

Burtraw, Dallas. 1996. Cost savings sans allowance trades? Evaluating the SO_2 emissions trading program to date. Discussion paper 95–30-REV, Resources for the Future.

Business Week. 1997. Silicon Valley: How it really works. 18–25 August: 64–147.

CACIWSS (Citizen Advisory Commitee for Industrial Water and Sewer Services). 1997. Final Report for Dominion Semiconductor Industrial Water and Sewer Facilities. Prince William County Board of Supervisors, 13 November.

California Department of Toxic Substances. 1994. Assessment of the Semiconductor Industry Source Reduction Planning Efforts. Office of Pollution Prevention and Technology Development.

Cambou, Bertrand F. 1996. PICO Frontier: The semiconductor industry is pursuing a great number of processes and devices to transform the IC to integrated systems on a chip. *Electronic Engineering Times Online* no. 934 (30 December). Available at http://www.techweb.com.

Capria, Antonella, ed. 1996. *A World Survey of Environmental Laws.* Giuffre Editore.

Chandrasekaran, Rajiv. 1996. Motorola postpones chip plant project. *Washington Post,* 16 November.

Chunn, Sherri. 1997. Intel boss urges end to red tape. *Albuquerque Journal,* 15 December.

Clinton, Bill. 1995. Remarks on Project XL. Old Executive Office Building, 3 November.

Clinton, Bill, and Al Gore. 1995. White House Policy on Reinventing Environmental Regulation. Executive Office of the President.

Cohen, Stephen S., and John Zysman. 1987. *Manufacturing Matters: The Myth of the Post-Industrial Economy.* Basic Books.

Cole, Henry S. 1996. The Final Project Agreement for the Intel XL Should Be Strengthened. H. S. Cole & Associates, Washington.

Coombs, Jim. 1996. Challenges in the Intel Stakeholder Process. The Alternative Path. Available at http://www.alt-path.com.

Corcoran, Elizabeth. 1997. Microsoft's Bill, on Capitol Hill. *Washington Post,* 5 June.

CRT (Campaign for Responsible Technology). 1996a. Coalition of Community, Environmental/Justice, and Labor Organizations Blast Clinton Administration "Sweetheart Deal" with Intel. Memo to members of CRT's advisory board, November.

CRT. 1996b. Newsletter, 6 March.

CRWQB (California Regional Water Quality Board, San Francisco Bay Region). 1995. Annual Report, Site Management System.

CSM (Chartered Semiconductor Manufacturing Inc.). 1997. Chartered Semiconductor Manufacturing Named Top Fab by Semiconductor International Magazine. Press release, 5 May. Available at http://www.csminc.com.

Cushman, John H., Jr. 1997. Virginia seen as undercutting US environmental rules. *New York Times,* 19 January.

Cyrix Corp. 1996. 1995 Annual report. Available at http://www.cyrix.com.

Dataquest. 1997. Top 10 Worldwide Semiconductor Market Share Estimates (1996).

Davies, J. Clarence, and Jan Mazurek. 1996. *Industry Incentives for Environmental Improvement: Evaluation of US Federal Initiatives.* Global Environmental Management Initiative.

Davies, J. Clarence, and Jan Mazurek. 1997. *Regulating Pollution: Does the US System Work?* Resources for the Future and Johns Hopkins University Press.

Davies, J. Clarence, and Jan Mazurek. 1998. *Pollution Control in the United States: Evaluating the System.* Resources for the Future and Johns Hopkins University Press.

DeJule, Ruth. 1998. New fab construction: Flexible facility architecture, safety, and energy usage concerns become paramount as costs soar. *Semiconductor International,* January: 81–84.

Dick, A. R. 1991. Learning by doing and dumping in the semiconductor industry. *Journal of Law and Economics* 40: 359–375.

Dickerson, R., H. B. Gray, and M. Y. Darensbourg. 1984. *Chemical Principles.* Benjamin/Cummings.

Dillon, Patricia S. 1994. Salvageability by design. *IEEE Spectrum* 31, no. 8: 18–21.

Dosi, Giovanni. 1988. Sources, procedures and microeconomic effects of innovation. *Journal of Economic Literature* 26: 1120–1171.

Economist. 1997. Silicon Valley survey and Deep in the heart of Texas. 29 March: 1–20.

EDB (Economic Development Board of Singapore). 1996a. More Semiconductor Companies Doing Multi-Activities. Available at http://www.sedb.com.

EDB. 1996b. Two New Schemes to Boost Semiconductor Industry. Available at http://www.sedb.com.

EIGNC (Electronics Industry Good Neighbor Campaign). 1997. *Sacred Waters: Life-Blood of Mother Earth.*

Electronic Business. 1987. New ventures in the semiconductor industry. 15 August: 46–64.

Ellram, L. M. 1990. The supplier selection decision in strategic partnerships. *Journal of Purchasing and Materials Management* 26, no. 4: 8–14.

Executive Office of the President. 1994. Report of the Economic and National Security Working Group on Lithography. Office of Science and Technology Policy.

Fehr-Snyder, Kerry. 1997. Valley chipmakers ruining water, say environmentalists. *Arizona Republic,* 8 August.

Ferguson, Sheila A. 1992. Influence of CBI Requirements on TSCA Implementation. Hampshire Research Associates, Inc.

Florida, Richard, and Martin Kenney. 1990. *The Breakthrough Illusion.* Basic Books.

Forester, Tom. 1993. *Silicon Samurai: How Japan Conquered the World's IT Industry.* Blackwell Business.

Fraust, Charles, Phillip L. Cornejo, R. Byron Davis, Edward R. Miroslaw, and Ilse Still. 1992. Environmental control in semiconductor manufacturing. *AT&T Technical Journal,* March-April: 19–25.

Freeman, Harry, Teresa Harten, Johnny Springer, Paul Randall, Mary Ann Curran, and Kenneth Stone. 1992. Industrial pollution prevention: A critical review. *Journal of the Air and Waste Management Association* 42, no. 5: 617–653.

Freemantle, Michael. 1995. Protein devices may increase computer speed and memory. *Chemical & Engineering News,* 22 May: 24–26.

Froines, John, Robert Gottlieb, Maureen Smith, and Pam Yates. 1995. Disassociating toxics policies: Occupational risks and product hazards. In *Reducing Toxics,* ed. R. Gottlieb. Island.

FSA (Fabless Semiconductor Association). 1997. Members of the FSA. Available at http://www.fsa.org.

Fuller, Brain. 1996. Executives: Semi story is sobering. *Electronic Engineering Times Interactive,* no. 884, 15 January. Available at http://www.techweb.com.

Garreau, Joel. 1991. *Edge City: Life on the New Frontier.* Doubleday.

Ginsberg, Beth S., and Cynthia Cummis. 1996. EPA's Project XL: A Paradigm for Promising Regulatory Reform. *Environmental Law Reporter* 26: 10057–10064.

Glaberson, William, and Julia Campbell. 1996. Ailing computer-chip workers blame chemicals, not chance. *New York Times,* 28 March.

Glasser, Lance. 1993. ARPA and the Environment. Presented at Electronics and Environment conference, Washington.

Gordon, Richard, and Joel Kreiger. 1994. Technological Change, Production Organization and Skill Formation in the US Machine Tool, Semiconductor and Auto Industries. Working paper 94-1. University of California, Santa Cruz: Center for the Study of Global Transformations.

Gottlieb, Robert. ed., 1995. *Reducing Toxics: A New Approach to Industrial Decisionmaking.* Island.

Greene, Stephen G. 1996. Foundation's shareholder activism: Grant makers use their status to persuade Intel to change its policy and share information with local communities, *Chronicle of Philanthropy.* 25 January.

Greer, Linda, and Christopher van Löben Sels. 1997. When Pollution Prevention Meets the Bottom Line. *Environmental Science and Technology* 31: 418–422.

Grove, Andrew S. 1991. Should the US abandon computer manufacturing? *Harvard Business Review,* September-October: 140–161.

Grove, Andrew S. 1996. *Only the Paranoid Survive: How to Exploit the Crisis Points That Challenge Every Company and Career.* Currency/Doubleday.

Gruber, Harald. 1992. Persistence of leadership in product innovation. *Journal of Industrial Economics* 40: 359–375.

Gruber, Harald. 1994. *Learning and Strategic Innovation: Theory and Evidence for the Semiconductor Industry.* North-Holland.

Gwennap, Linley. 1996. Intel's geography lesson. *Microprocessor Report* 10, no. 3: 2.

Hamilton, Martha M. 1996. The economy gets a new taxonomy: Analysts welcome a SIC transit to '90s realities. *Washington Post,* 28 June.

Hammond, J. L., and Barbara Hammond. 1919. *The Skilled Labourer, 1760–1832.* Longmans.

Harrison, Bennett, and Barry Bluestone. 1982. *The Deindustrialization of America.* Basic Books.

Harrison, Myron. 1992. Semiconductor manufacturing hazards. In *Hazardous Materials Technology,* ed. J. B. Sullivan Jr. and G. Kreeiger. Williams & Wilkins.

Hatcher, Julia A. 1994. Comments by Intel Corporation on the proposed Amendments to the Criteria for Interim Approval of Title V programs before the United States Environmental Protection Agency. 28 September.

Hawes, Amanda. Not dated. Workplace Hazards for High Tech Workers. Santa Clara Center for Occupational Safety and Health.

Hayashi, Alden M. 1988. The new shell game: Where is the US chip really made? *Electronic Business,* 1 March: 36–38.

Herman, R., S. A. Ardekani, and J. H. Ausubel. 1989. Dematerialization. In *Technology and Environment* edited by J. H. Ausebel and H. D. Sladovich. National Academy Press.

Holcomb, Mark A. 1997. Conversation with J. Mazurek, 17 December.

Horvath, Arpad, Chris Hendrickson, Lester Lave, Francis McMichael, and Tse-Sung Wu. 1995. Toxic emissions indices for green design and inventory. *Environmental Science & Technology* 29, no. 2: 86A–88A.

Hossfeld, Karen. 1995. Down and out in Silicon Valley. *Amagazine,* October-November: 30–35. Available at http://www.amagazine.com.

Hoover's Online. 1998. Industry snapshot: Semicondutors. Available at http://www.hoovers.com.

Hsu, Spencer S. 1997. EPA investigating Virginia agency: Warning of takeover. *Washington Post,* 15 July.

Hubner, John. 1997. Cause and effect and cancer. *West Magazine (San Jose Mercury News),* 25 May.

Hutcheson, G. Dan, and J. D. Hutcheson. 1996. Technology and Economics in the Semiconductor Industry. VLSI Research.

IBM (International Business Machines Corp.). 1996. IBM and the Environment. Available at http://www.ibm.com.

IBM. 1997. Electronic mail correspondence with Edan Dionne, Corporate Environmental Affairs, 14 October.

ICE (Integrated Circuit Engineering, Phoenix). 1993. Fab status report.

Inside EPA. 1996a. EPA developing plan for site-specific rules for XL projects. *Inside EPA,* 12 April: 1–2.

Inside EPA. 1996b. EPA is considering possible performance tests for XL Projects. *Inside EPA,* 16 August: 5–6.

Inside EPA. 1997. Industry groups abandon EPA's Common Sense Initiative. *Inside EPA,* 31 January: 1–10.

Intel Corp. 1995. Environmental Health and Safety at Intel 1994.

Intel Corp. 1996a. Evolution of Environmental Management at Intel. Report presented at Fab 12 Project XL meeting, 24 January.

Intel Corp. 1996b. Minutes of Fab 12 Project XL meeting, 24–25 January. Available at http://www.intel.com.

Isaacson, Walter. 1997. Driven by the passion of Intel's Andrew Grove. *Time,* 29 December.

ISO (International Organization for Standardization). 1995. Environmental management systems—specification with guidance for use—draft International Standard ISO 14001.

Kirkpatrick, David. 1997. Intel's amazing profit machine. *Fortune* 17, February: 60–72.

Kirschner, Elisabeth M. 1995. Chemical industry modernizes aging plants for safety and efficiency. *Chemical & Engineering News,* 10 July.

Kneese, Allen V., Robert U. Ayres, and Ralph C. d'Arge. 1970. *Economics and the Environment: A Materials Balance Approach.* Johns Hopkins University Press for Resources for the Future.

Kogut, B., and D. Kim. 1991. Strategic Alliances of Semiconductor Firms. Unpublished report to Dataquest, San Jose.

Krugman, Paul. 1991. *Geography and Trade.* MIT Press.

Krugman, Paul. 1996. *Pop Internationalism.* MIT Press.

Krupnick, Alan J., Allen Blackman, and James Boyd. 1998. Towards an economic theory of site-specific environmental regulation: Implications for policy. Resources for the Future.

LaDou, Joseph. 1986. Health issues in the microelectronics industry. *Journal of Occupational Medicine* 1: 1–11.

LaDou, Joseph, and Timothy Rohm. 1998. The international electronics industry. In *Environmental and Occupational Medicine,* ed. W. Rom. Lippincott-Raven.

Lamond, A., and R. Wilson. 1984. *The Competitive Status of the US Electronics Industry.* National Academy Press.

LaPedus, M. 1996. The next wave: A new generation of chip makers target foundries and DRAMs to compete with neighbors Korea and Japan. *Electronic Buyer's Guide,* 2 January.

Lavelle, Marianne. 1996. Bending the rules: Project lets companies achieve pollution goals by "breaking the law." *National Law Journal* 18, June: 1.

Leopold, George. 1994. Gore unveils semiconductor initiative. *Electronic Engineering Times,* no. 787, 7 March.

Lewis, Sanford. 1997. Greening of the Electronics Industry: Prospects and Dilemmas for Stakeholder Participation. Pollution Prevention Education and Research Center, University of California. Los Angeles.

Malone, Michael S. 1996. Chips triumphant. *Forbes ASAP.* 26 February, 53–78.

Markoff, John. 1997. A gold rush from software animates the Silicon Valley. *New York Times,* 13 January.

Matusow, David. 1996. Thoughts on Public Involvement in Project XL. Memo on file with J. Mazurek.

Maxie, E. 1994. Supplier performance and the environment. Presented at IEEE International Symposium on Electronics and the Environment, San Francisco.

Maxwell, James, Julia Bucknall, and John Ehrenfeld. 1993. The Response of the Electronics Industry. MIT Norwegian Chlorine Study.

Mazurek, Janice V. 1994. How Fabulous Fablessness: Environmental Planning Implications of Economic Restructuring in the Silicon Valley Semiconductor Industry. Master's thesis, University of California, Los Angeles.

MCC (Microelectronics and Computer Technology Corp.). 1993. Environmental Consciousness: A Strategic Competitiveness Issue for the Electronics and Computer Industry.

Miller, Roger E., and Marcel Cote. 1987. *Growing the Next Silicon Valley: A Guide for Successful Regional Planning.* Lexington Books.

Milliman, Kevin. 1995. ISO 14000 and US Manufacturing. Center for Risk Management, Resources for the Future.

Ministry of the Environment, Singapore. 1997. Pollution Control Functions. Available at http://www.gov.

Mohin, Timothy. 1996. Telephone conversation with J. Mazurek, 11 August. Mohin is a Government Affairs Manager in Intel's Environmental Health and Safety department.

Mohin, Timothy. 1997. The alternative compliance model: A bridge to the future of environmental management. *Environmental Law Reporter* 27: 10345–10356.

Morris, P. R. 1990. *History of the World Semiconductor Industry.* Peter Peregrinus.

NAPA (National Academy of Public Administration). 1994. Common Sense Initiative. Washington, D.C. : NAPA.

NAPA. 1995. *Setting Priorities, Getting Results: A New Direction for EPA.*

NAPA. 1997. Case study 1: Excellence, leadership, and the Intel Corporation: A study of EPA's Project XL. In *Resolving the Paradox of Environmental Protection.*

National Semiconductor. 1995. Annual report. Available at http://www.national.com.

National Semiconductor. 1996. Worldwide Manufacturing Locations: Greenock, Scotland. Available at http://www.national.com.

National Semiconductor. 1997. List of worldwide manufacturing facilities. Available at http://www.national.com.

Navin-Chandra, D. 1991. Design for environmentability. Presented at Design Theory and Methodology conference, American Society of Mechanical Engineers, Miami.

Nelson, Richard R., and Sidney G. Winter. 1982. *An Evolutionary Theory of Economic Change.* Harvard University Press.

NRDC (Natural Resources Defense Council). 1996a. Comments on the Intel Project XL, 3 July.

NRDC. 1996b. Supplemental comments on the draft Final Project Agreement (FPA) 25 July, 14 August, 13 September.

NSC (National Safety Council). 1997. Electronic Product Recovery & Recycling Conference EPR2. Summary Report. Environmental Health Center, Washington.

ODEQ (Oregon Department of Environmental Quality). 1994. Oregon Title-V operating permit. Northwest Region. Portland. Permit Number 34–2681.

O'Harrow, Robert Jr. 1997. Save to disk? Yes, and make a printout, too: In computer age, we still want hard copies. *Washington Post,* 2 November.

Parks, Bob. 1997. After life: Where computers go to die. *Wired 5,* no. 7: 146. Available at http://www.wired.com.

Pedersen, William F., Jr. 1995. Can site-specific pollution control plans furnish an alternative to the current regulatory system and a bridge to the new one? *Environmental Law Review* 25: 10486–10490.

Piore, Michael and Charles F. Sabel. 1984. *The Second Industrial Divide*. Basic Books.

Plazola, Carlos. 1997. The Globalization of High Tech: Environmental Injustices Plague Industry. Silicon Valley Toxics Coalition memo, on file with J. Mazurek.

PNPPC (Pacific Northwest Pollution Prevention Center). 1995. Oregon Intel Title V permit.

Port, Otis. 1997. Gordon Moore's crystal ball. *Business Week*, 23 June: 120.

Radford, Jeff. 1993. Intel air quality violations will bring fines, state says. *Corrales Comment*, 23 October.

Rappaport, Andrew S. 1992. Fabless forever. *Electronic Engineering Times*, no. 684.

Rice, Valerie. 1987. The Upstart Start-Ups. *Electronic Business*, 15 August: 46–61.

Rivera, José A. 1996. The Acequias of New Mexico and the Public Welfare. School of Public Administration, University of New Mexico.

Robinson, Gail. 1996. Lattices self-assemble. *Electronic Engineering Times Online* no. 912 (29 July). Available at http://www.techweb.com.

Sarkis, Joseph, Nicole M. Darnall, Gerald I. Nehman, and John W. Priest. 1995. The role of supply chain management within the industrial ecosystem. Presented at IEEE International Symposium on Electronics and the Environment, Orlando.

Saxenian, AnnaLee. 1981. Silicon Chips and Spatial Structure: The Industrial Basis of Urbanization in Santa Clara County. Working paper 345, Institute of Urban and Regional Development, Berkeley.

Saxenian, AnnaLee. 1994. *Regional Advantage: Culture and Competition in Silicon Valley and Route 128*. Harvard University Press.

Schierow, Linda-Jo. 1994. Comparison of environmental risk provisions in the 103d Congress. *Risk* 5: 253–283.

Schumpeter, Joseph A. 1934. *The Theory of Economic Development*. Oxford University Press.

Scott, Allen J. 1988. The global assembly operations of US semiconductor firms: A geographical analysis. *Environment and Planning A* 20: 1047–1067.

Scott, Allen J., and David P. Angel. 1987. The US semiconductor industry: A locational analysis. *Environment and Planning A* 19: 875–912.

SEEQ Technology Inc. 1991. Annual report.

SEEQ Technology Inc. 1996. Corporate Fact Sheet. Available at http://www.seeq.com.

SEEQ Technology Inc. 1997a. 1996 Annual report.

SEEQ Technology Inc. 1997b. Electronic mail correspondence with James Middleton, Director of Operations, 29 October.

Sematech. 1997. Discover a new world of opportunity. Available at http://www.4chipjobs.com.

SEMI (Semiconductor Equipment and Materials International). 1993. Safety Guidelines for Semiconductor Manufacturing Equipment (SEMI S2–93).

SEMI. 1994. Fabs of the Southwest.

Sheppard, Bill F. 1995. Testimony on behalf of Intel Corporation regarding the Operating Permit Program Requirements of the Clean Air Act. House Subcommittee on Oversight and Investigations of the Committee on Commerce, 103rd Congress, 18 May.

Sherry, Susan. 1985. High Tech and Toxics: A Guide for Local Communities. Golden Empire Health Systems Agency.

SIA (Semiconductor Industry Association). 1994. The National Technology Roadmap for Semiconductors.

SIA. 1997. Worker Safety & Environment. Available at http://www.semichips.org.

SIA. 1998. Global Semiconductor Sales to Decline 1.8 Percent in 1998. Available at http://www.semichips.org.

Skrzycki, Cindy. 1996. Critics see a playground for polluters in EPA's XL plan. *Washington Post*, 24 January.

Skrzycki, Cindy. 1997. Some state environmental chiefs want EPA off the stage. *Washington Post*, 28 June.

Slater, Michael. 1996a. Electronic mail correspondence with J. Mazurek, 20 September.

Slater, Michael. 1996b. Intel's new world view: Market growth, not competition, is top concern. *Microprocessor Report* 10, no. 2: 1. Available at http://www.chipanalyst.com.

Smith, Maureen. 1997. *The US Paper Industry and Sustainable Production: An Argument for Restructuring*. MIT Press.

Smith, Rebecca. 1994. Why Intel jilted the Golden State. *San Jose Mercury News*, 20 July: A1–A17.

Smith, Ted. 1992. Celebrating Ten Years of Progress. Silicon Valley Toxics Coalition.

Smith, Ted, and Phil Woodward. 1992. The legacy of high tech: The toxic lifecycle of computer manufacturing. Silicon Valley Toxics Coalition.

Stanaback, Steve. 1997. Union County pilot program contends with "oddball waste stream." *Product Stewardship Advisor* 1, no. 1: 8–10.

State Environmental Monitor. 1997. USEPA drafts Project XL rules to define "superior" performance. *State Environmental Monitor*, 6 January.

Storper, Michael, and Richard Walker. 1989. *The Capitalist Imperative: Technology, Territory and Industrial Growth*. Blackwell.

SVTC (Silicon Valley Toxics Coalition). 1995. Did you know? *Silicon Valley Toxics Action* 13, no. 4: 8.

SVTC. 1997. Milestones: Silicon Valley Toxics Coalition history 1982–1996. *Silicon Valley Toxics Action* 15, no. 1: 7.

SWOP (SouthWest Organizing Project). 1995. *Intel Inside New Mexico: A Case Study of Environmental and Economic Injustice*.

Texas Department of Commerce. 1996. Texas Department of Commerce Programs Instrumental in Samsung Announcement. Press release, 16 January.

Thurow, Lester. 1992. *Head to Head: The Coming Economic Battle Among Japan, Europe, and America*. Morrow.

TSMC (Taiwan Semiconductor Manufacturing Company). 1998a. About TSMC: Current Status. Available at http://www.tsmc.com.

TSMC. 1998b. Certified ISO-14001. Available at http://www.tsmc.com.

Tyson, Laura D'Andrea. 1992. *Who's Bashing Whom: Trade Conflict in High Technology Industries*. Institute for International Economics.

United Church of Christ. 1987. Toxic Wastes and Race Revisited.

US DoC (Department of Commerce). 1994. County business patterns. In US Summary Data (Bureau of the Census). Available at http://www.census.gov.

US DoC. 1997a. 1994 County Business Patterns for Santa Clara, CA (Bureau of the Census). Available at http://www.census.gov.

US DoC. 1997b. Current Industrial Reports (Bureau of the Census). MA36 Series.

US EPA (Environmental Protection Agency). Not dated. An introduction to EPA's Design for the Environment Program. Office of Pollution, Prevention and Toxics.

US EPA. 1990. Toxics in the Community: National and Local Perspectives. Office of Pollution Prevention and Toxics.

US EPA. 1992. Energy Star Computers Program. Global Change Division.

US EPA. 1994a. Common Sense Initiative. Administrator's update. Office of the Administrator, 29 July.

US EPA. 1994b. Environmental Justice Annual Report: Focusing on Environmental Protection for All People. Office of Environmental Justice.

US EPA. 1995a. Project XL Proposals for Facilities, Sectors, and Government Agencies. Office of Policy, Planning and Evaluation.

US EPA. 1995b. EPA Office of Compliance sector notebook project: Profile of the electronics and computer industry. Office of Compliance.

US EPA. 1995c. A facility-based alternative system of environmental protection. Computers and Electronics Subcommittee. Draft. 29 January. Office of Pollution, Prevention, and Toxics.

US EPA. 1996a. Draft air permit conditions. Office of Policy, Planning and Evaluation.

US EPA. 1996b. Letter from Deputy Administrator John Wise, EPA Region IX to Karen Heidel, deputy director of the Arizona Department of Environmental Quality, and Al Brown Maricopa County Environmental Services Department, 6 November.

US EPA. 1996c. EPA will use the following criteria to evaluate Project XL proposals. Office of Policy, Planning and Evaluation.

US EPA. 1996d. Principles for development of Project XL Final Project Agreements. Office of Policy, Planning and Evaluation.

US EPA. 1996e. Project XL: Final Project Agreement for the Intel Corporation Ocotillo Site Project XL. Office of Policy, Planning and Evaluation.

US EPA. 1996f. Response to Comments on Intel's Final Project Agreement. Office of Policy, Planning and Evaluation.

US EPA. 1996g. 1994 Toxics Release Inventory: Public Data Release. Office of Pollution Prevention and Toxics.

US EPA. 1996h. Letter from the Washington Office on Environmental Justice to the Honorable Carol M. Browner, Administrator, US Environmental Protection Agency dated 5 February. Office of the Administrator.

US EPA. 1997a. New Directions: A Report on Regulatory Reinvention. Office of the Administrator.

US EPA. 1997b. Toxics Release Inventory: Public data release. Office of Pollution, Prevention and Toxics.

US EPA. 1998. XL Project Information. Available at http://199.223.29.233/ProjectXL/xl_home.nsf.

US GAO (General Accounting Office). 1990. EPA's Chemical Testing Program Has Made Little Progress. GAO/RCED-90–112.

US SEC (Securities and Exchange Commission). 1997. Edgar database 10K Reports. Available at http://www.sec.gov.

Van Zant, Peter. 1990. *Microchip Fabrication: A Practical Guide to Semiconductor Processing*. McGraw-Hill.

VEDP (Virginia Economic Development Partnership). 1997. The Virginia file: Executive summary.

Vonnegut, Kurt, Jr. 1952. *Player Piano*. Scribner.

Weber, Samuel. 1991. A new endangered species: Not many US semiconductor houses can afford to manufacture. Is competitiveness at risk? *Electronics*, July: 36–43.

White, Robert. 1996. Supes order up H20 test: River salt's a concern. *Fairfax Journal*, 19 November.

Williams, Michael E., David G. Baldwin, and Paul C. Manz. 1995. *Semiconductor Industrial Hygiene Handbook: Monitoring, Ventilation, Equipment and Ergonomics*. Noyes.

Williamson, Oliver E. 1975. *Markets and Hierarchies: Analysis and Antitrust Implications.* Free Press.

Wood, Maureen. 1997. Telephone conversation with J. Mazurek, 28 October.

Wood, Maureen. 1998. Report on status of Dominion Semiconductor plant (conference call on wafer fabrication in Virginia, sponsored by Silicon Valley Toxics Coalition, 27 February).

Zhang, Jing Yang. 1997. Reflections on international environmentalism. *Silicon Valley Toxics Action* 15, no. 3: 3.

Index